CREATINE

BIOSYNTHESIS, THERAPEUTIC USES AND PHYSIOLOGICAL EFFECTS OF SUPPLEMENTATION

BIOCHEMISTRY RESEARCH TRENDS

PHYSIOLOGY - LABORATORY AND CLINICAL RESEARCH

CREATINE

BIOSYNTHESIS, THERAPEUTIC USES AND PHYSIOLOGICAL EFFECTS OF SUPPLEMENTATION

FERNANDO D'CRUZ

AND

VICTOR RIBEIRO

EDITORS

New York

For permission to use material from this book please contact us:
Telephone 631-231-7269; Fax 631-231-8175
Web Site: http://www.novapublishers.com

NOTICE TO THE READER

Additional color graphics may be available in the e-book version of this book.

Library of Congress Cataloging-in-Publication Data

ISBN: 978-1-62948-305-4

Library of Congress Control Number: 2013950461

Published by Nova Science Publishers, Inc. † New York

Contents

Preface

Creatine is an organic acid that contributes to the energy supply to the skeletal muscle. The basic substrates for creatine biosynthesis in the human body are semi-essential amino acids L-arginine, glycine and methionine. In this book, the authors discuss the biosynthesis, therapeutic uses and physiological effects of creatine supplementation. Topics include the role of creatine in the pathophysiology of depression and the possible mechanisms underlying its antidepressant effect; creatine treatment and positive effects on muscle performance, muscle mass gain, and the persistence and improvement of the quality of life in patients with chronic diseases; muscle ergogenic effects of creatine supplementation in resistance exercise training; experimental evidence that creatine supplementation during pregnancy is protective for the neonate; and creatine metabolism and role in sports physiology.

Chapter I – Depression is a common mood disorder associated with high rates of morbidity and mortality and characterized by a broad range of symptoms, including altered mood and cognitive functions, loss of interest or pleasure, feelings of guilt or low self-worth, disturbed sleep or appetite, low energy and recurrent thoughts of death or suicide. Although a number of new medications and therapies have been used for the management of this disorder, most of them were based on serendipitous discoveries. Consequently, the management of depression remains little changed over the last decades. However, accumulated evidence has suggested that an alteration in the brain energy metabolism (particularly a hypoactive prefrontal cerebral energy metabolism) may be implicated in the pathophysiology of this disease. It is therefore possible that agents that improve cellular bioenergetics may exert beneficial effects for treating depression. In this context, emerges creatine. The creatine kinase/phosphocreatine system plays a key role in the brain energy homeostasis by maintaining high ADP levels at the site in mitochondria at which ATP is generated and low levels at the site of ATP utilization. This chapter reviews the available findings that suggest that this compound exerts a role in the pathophysiology of depression and the possible mechanisms underlying its antidepressant effect.

Chapter II – Creatine is an organic acid that contribute to the energy supply to the skeletal muscle. The basic substrates for the creatine biosynthesis in the human body are semi-essential amino acids L-arginine, glycine and methionine. The creatine is synthesized in the liver and kidneys and partially can be substituted from food such as fish or meat. It is an essential source of muscle energy, which is used mainly by athletes to improve short-term performance. The creatine is one of the best-selling nutritional supplements. Effects of the creatine on increasing muscle strength was also tested in pathological conditions and diseases such as injuries or surgeries associated with prolonged immobilization, in patients with muscle weakness associated diseases, in patients with increased fatigue, in some neuromuscular diseases, in patients with muscular dystrophy, amyotrophic lateral sclerosis, chronic obstructive pulmonary disease or in patients with chronic heart failure. The creatine can be applied in various forms - creatine monohydrate, creatine phosphate, creatine maleate, creatine citrate etc. Based on available data the authors can conclude that the effect depends on the pharmacokinetic parameters of the forms of the creatine used for the supplementation. After per oral supplementation the creatine is rapidly and reversibly phosphorylated in the muscles. This reaction is catalyzed by creatine kinase, which produces creatine phosphate as a basic source of ATP resynthesis. ATP is subsequently used as a basic energy source. Use of the creatine is not introduced yet to standard practice, however some of the experimental and clinical studies indicate that supplementation of the creatine improves function of skeletal muscle in some diseases (such as COPD), and consequently the quality of life of affected patients. The creatine supplementation can replace anabolic steroids or nutritional supplements in conditions associated with malnutrition, such as cancers or COPD. Very interesting areas represent neuroprotective effects of creatine, effects on cognitive and mental functions, influence on the metabolism of carbohydrates and fats. Currently, the use of the creatine is very popular especially among young people and athletes in order to increase muscle strength, muscle growth and weight loss. The creatine may also, as mentioned above, improve the performance especially in relation to short-term sports activity. The results of the creatine supplementation in the pathological conditions remain controversial. The studies reported positive results but the effects of the creatine often depend on many factors like comorbidities and individual metabolic differences. The review summarizes the current therapeutic possibilities of the creatine as well as the results of recent research activity in the creatine field.

Chapter III – Creatine in its monohydrated form (CrM) is on the top of the list among athletes seeking ergogenic supplements for increasing muscle mass and/or strength. It is proposed that CrM supplementation induces cell swelling which would activate downstream cell-volume signaling cascades and affect muscle-cell metabolism. The up regulated protein involved in osmosensing would result in increases in maximal muscle strength and power, fat-free mass, total body water and body weight. These would happen independently of training but overall CrM supplementation actions usually potentiate concurrent heavy-resistance training effects on muscle. CrM has also a major role on post-exercise induced damage by slowing down the degeneration phase and accelerating the regeneration phase after high-intensive exercises. As such CrM may provide protective effects via increased phosphocreatine synthesis recovering the ATP levels and maintaining the Ca^{2+}-ATPase pump. Such effects would keep the Ca^{2+} inside the sarcoplasmatic reticulum avoiding Ca^{2+} dependent proteases activation, protein breakdown and higher cell membrane permeability (intracellular components leakage). Traditional CrM supplementation protocols include phases of loading and maintenance. Co-ingestion of CrM with carbohydrate and protein may enhance Cr uptake through insulin action. Although there may be various side effects of Cr supplementation the only clinically significant side effect scientifically proven has been body weight gain. Nevertheless, it is recommended to avoid high dosages of Cr during periods of increased thermal stress as those under high ambient temperature/humidity. Thus Cr supplementation can be considered a safe, ethical, legal, inexpensive and effective nutritional intervention particularly when consumed in conjunction with a resistance training protocol.

Chapter IV – The ergogenic and neuroprotective properties of creatine are now well recognised, and creatine supplementation is used widely by healthy adults and as a therapy for patients with neurological and musculo-skeletal conditions where depletion of ATP might be prevented by increasing the intracellular creatine pool. Babies can be born under conditions where poor placental perfusion, intrauterine infection, or depressed respiratory drive result in oxygen deprivation, increasing the risk of damage arising in the brain and other key organs of the neonate due to the inability of cells to rapidly replenish ATP via the creatine kinase pathway. This chapter reviews recent experimental work in pregnant and newborn animals that shows the benefits of creatine supplementation for survival after birth when oxygen and nutrient supply may limit cellular respiration, providing the basis for further consideration of the use of creatine in human pregnancy.

Chapter V – Methyl guanidine-acetic acid, commonly named creatine (Cr), is a naturally occurring amino acid provided by diet or synthesized endogenously, primarily in liver, pancreas and kidneys. Cr biosynthesis involves two enzymes; arginine glycine amidinotransferase that converts arginine and glycine to ornithine and guanidinoacetate, and S-adenosyl-L-methionine:N-guanidinoacetate methyltransferase that converts guanidinoacetate to Cr. Cr is taken up from blood into Cr-requiring tissues against a concentration gradient by a specific transporter (SLC6A8). Approximately 95% of Cr is stored in skeletal muscle and the remaining 5% is stored in the heart, brain and testes. Cr and its derivative phosphocreatine (PCr) play an essential role in energy storage and transmission in most tissues; predominant role being played in skeletal muscle and brain. In athletes, PCr is particularly important during explosive activities when a high rate of energy release is required. PCr serves as a source of high energy phosphates and increasing PCr muscle stores might have a significant effect on physical performance, especially during anaerobic intense exercises. Several researchers have evaluated the potential ergrogenic value of Cr. It has been shown that Cr supplementation increases total Cr content in skeletal muscle, particularly in individuals with lowest basal Cr content. Numerous studies have examined the effect of Cr supplementation in athletes. Most studies have shown that Cr supplementation is accompanied by a significant increase in body weight, with increase of lean mass and decrease of fat mass. This may be due to stimulation of muscle protein synthesis or water retention, a point that is particularly relevant for athletes in explosive sports disciplines. The results support the proposal that Cr supplementation improves performance particularly during high intensity exercise, as well as resistance training. However, other studies failed to show a significant effect of Cr supplementation. Issues such as the dose and duration of supplementation, exercise protocol, and sports disciplines may have contributed to these discrepancies. There are some anecdotal reports on potentially adverse effects of Cr supplementation such as muscle cramps, gastrointestinal complaints, liver dysfunction, and kidney impairment. It appears that Cr supplementation at the correct dose is not harmful to healthy athletes, but it must be used with precaution in individuals at risk. Finally, it seems reasonable to monitor liver and kidney functions in individuals consuming Cr supplements.

In: Creatine
Editors: F. D'Cruz and V. Ribeiro

ISBN: 978-1-62948-305-4
© 2013 Nova Science Publishers, Inc.

Chapter I

The Role of Creatine in the Pathophysiology of Depression and the Possible Mechanisms Underlying its Antidepressant Effect

***Mauricio P. Cunha, Francis L. Pazini,
Ágatha Oliveira and Ana Lúcia S. Rodrigues***
Department of Biochemistry, Center of Biological Sciences,
Universidade Federal de Santa Catarina, Florianópolis, SC, Brazil

Abstract

Depression is a common mood disorder associated with high rates of morbidity and mortality and characterized by a broad range of symptoms, including altered mood and cognitive functions, loss of interest or pleasure, feelings of guilt or low self-worth, disturbed sleep or appetite, low energy and recurrent thoughts of death or suicide. Although a number of new medications and therapies have been used for the management of this disorder, most of them were based on serendipitous discoveries. Consequently, the management of depression remains little changed over the last decades. However, accumulated evidence has suggested that an alteration in the brain energy metabolism (particularly a hypoactive prefrontal cerebral energy metabolism) may be implicated in the pathophysiology of this disease. It is therefore possible that agents that improve cellular bioenergetics may exert beneficial effects for treating depression. In this context, emerges creatine. The creatine kinase/phosphocreatine system plays a key role in the brain energy homeostasis by maintaining high ADP levels at the site in mitochondria at which ATP is generated and low levels at the site of ATP utilization. This chapter reviews the available findings that suggest that this compound exerts a role in the pathophysiology of depression and the possible mechanisms underlying its antidepressant effect.

List of Abbreviations

ADP	adenosine diphosphate
AGAT	L-arginine glycine amidine transferase
ATP	adenosine triphosphate
BCK	cytosolic brain-type creatine kinase
CaMK-2	Ca^{2+}/calmodulin-dependent protein kinase II
CNS	central nervous system
FST	forced swimming test
GAMT	guanidineacetate methyltransferase
GPx	glutathione peroxidase
GR	glutathione reductase
GSH	glutathione
GSSG	glutathione disulfide
HDRS	Hamilton Depression Rating Scale
$5\text{-}HT_{1A}$	5-hydroxytryptamine 1_A
MEK	mitogen-activated protein/extracellular signal-regulated kinase
NADPH	nicotinamide adenine dinucleotide phosphate
NMDA	N-Methyl-D-aspartate
NO	nitric oxide
Nrf2	nuclear factor erythroid 2-related factor 2
PI3K	phosphatidylinositol-3 kinase
PKA	protein kinase A
PKC	protein kinase C
ROS	reactive oxygen specie
SNRI	selective noradrenaline reuptake inhibitor
SSRI	selective serotonin reuptake inhibitor

TST   tail suspension test
TNF-α   tumor necrosis factor-α
uMtCK   ubiquitous mitochondrial creatine kinase

Introduction

Psychiatric disorders affect approximately 450 million people worldwide (World Health Organization, 2009). Among mood disorders, depression is the most prevalent psychiatric disorders in the population, with a lifetime prevalence of approximately 20% (Wong and Licinio, 2001; Nestler et al., 2002). Approximately one in six men and one in four women will experience a depressive episode throughout life (Kessler et al., 2005; Holden, 2005). Depression is a disorder characterized by a broad range of symptoms, including altered mood and cognitive functions, loss of interest or pleasure, feelings of guilt or low self-worth, disturbed sleep or appetite, low energy and recurrent thoughts of death or suicide (Wong and Licinio, 2001). In contrast with the normal experiences of sadness, clinical depression is a chronic disease that can interfere significantly in the individual's life quality (Nakajima et al., 2010).

Despite the importance of depression, the etiology of this disorder is not well established (Wong and Licinio, 2001; Elhwuegi, 2004), but it is known to be associated with genetic and environmental (nongenetic) factors (Krishnan and Nestler, 2008).

Epidemiologic studies show that about 30% to 40% of the risk for depression is genetic (Caspi et al., 2010; Dick et al., 2010). Nongenetic factors as stress and emotional trauma, viral infections (e.g., Borna virus), and even random processes during brain development have been also implicated in its etiology (Fava and Kendler, 2000; Wong and Licinio, 2001; Berton and Nestler, 2006; Nemeroff, 2007; Pittenger and Duman, 2008). Moreover, depression occurs in the context of innumerable medical conditions such as endocrine disturbances (hyper or hypocortisolemia, hyper or hypothyroidism), neurodegenerative diseases, traumatic brain injury, cancer, asthma, diabetes and stroke (Nemeroff, 2007; McEwen et al., 2012; Morrison et al., 2012).

Initially, the antidepressant therapy was based on the inhibition of serotonin or noradrenaline reuptake transporters by the tricyclic antidepressants, and inhibition of monoamine oxidase (the major catabolic enzyme of monoamines) activity by monoamine oxidase inhibitors (Frazer, 1997).

These discoveries enabled the development of numerous second generation medications (e.g., selective serotonin reuptake inhibitors (SSRIs) and selective noradrenaline reuptake inhibitors (SNRIs) which are widely used today (Schmidt and Duman, 2007). Indeed, much depression research was based on the notion that understanding how these treatments work would reveal new insight into its causes (Trivedi, 2006). The treatment of depression has limited success mainly due to the delayed onset of therapeutic effect, partial or no responses attained by the patients, besides poor tolerability, high cost, and stigma associated with its use (Masand, 2003; Usala et al., 2008; Young et al., 2009). Indeed, the monoamine-based antidepressants remain the first line of therapy for depression, but their long therapeutic delays, low (about 30%) remission rates and their adverse effects have encouraged the search for more effective agents (Krishnan and Nestler, 2008; Rogoz et al., 2008; Covington et al., 2010).

Despite the heavy emphasis that has been given to the "monoaminergic hypothesis of depression" to explain the etiology of this disorder, several studies have shown that it is also associated with the metabolic system. A decrease in creatine and phosphocreatine levels, as well as a decrease in the creatine kinase activity and creatine transporter density have been reported in the brain of depressive patients and in rodents submitted to models of depression (Lugenbiel et al., 2010; Allen, 2012). In addition, preclinical studies have demonstrated that creatine supplementation exerts antidepressant effects that are associated with monoaminergic neurotransmission modulation (Cunha et al., 2012, 2013a,b). Clinical studies support the results obtained with experimental animals, indicating that creatine exhibits antidepressant properties (Kondo et al., 2011; Lyoo et al., 2012).

Taking into account that depressive disorder is associated with neurochemical alterations that can be counteracted by creatine and considering its neuroprotective and antidepressant effects reported both in preclinical and clinical studies; it is tempting to propose that this compound may be a novel strategy for the treatment of depression. Therefore, in this chapter, we provide an overview of the literature supporting the notion that creatine is implicated in the pathophysiology of depression and its potential antidepressant beneficial effects.

Neurobiological Basis of Depression

Monoaminergic System

Over the past 60 years, the role of monoaminergic system dysfunction has been studied in the pathophysiology and treatment of depression (Schildkraut, 1974). The "monoaminergic hypothesis of depression" postulates that in depressive patients there is a decrease on serotonin, noradrenaline and dopamine levels in central nervous system (CNS), as well as deficiencies in the amount and/or sensitivity of receptors for these neurotransmitters (Mann et al., 1996; Wong and Licinio, 2001). In line with this, the majority of drugs available in the market today modulate these systems. SSRIs, SNRIs and dopamine reuptake inhibitors block pre-synaptic transporters of these monoamines, increasing the levels of these biogenic amines in the synaptic cleft.

A monoaminergic receptor involved in the pathophysiology of depression is the 5-HT_{1A} receptor. This receptor has been associated with mood alterations, and in the mechanism of action of various classes of antidepressant drugs, including tricyclics, SSRIs and monoamine oxidase inhibitors (Albert, 2012; Hensler, 2002). The blockade of 5-HT_{1A} autoreceptors enhances presynaptic serotonergic function by preventing self-inhibitory actions of serotonin on the firing and release of serotonergic neurons (Artigas et al., 1994, 1996) and is associated with antidepressant response (Zomkowski et al., 2004, Kaster et al., 2005; Brocardo et al., 2008). Furthermore, the postsynaptic 5-HT_{1A} heteroreceptors mediate serotonergic actions on target neurons and have been implicated in regulation of mood, and stress responses (Albert, 2012).

In addition, several antidepressants have been found to increase α_1-adrenoceptors function (Holsboer and Barden, 1996), since they are able to increase the density of α_1-adrenoceptors in the hippocampus and cerebral cortex of mice and rats (Rehavi et al., 1980; Deupree et al., 2007) and the α_1-adrenoceptor agonist affinity in the cerebral cortex (Nowak and Przegaliński, 1988; Klimek et al., 1991). Moreover, the α_1-adrenoceptor blockade in the CNS induces depression-related behavior in the tail suspension test (TST), a test predictive of antidepressant activity (Stone and Quartermain, 1999).

The dopamine D_1 receptor activation has been associated with antidepressant responses of monoamine oxidase B inhibitors, serotonin reuptake inhibitors, tricyclics and the dopamine reuptake inhibitors in predictive tests of antidepressant action (Gambarana et al., 1995; Renard et al., 2001; Shimazu et al., 2005; Binfaré et al., 2010; Melo et al., 2011). Another receptor involved in the antidepressant response is dopamine D_2 receptor. Accordingly, the dopamine D_2 receptor antagonist sulpiride prevented the antidepressant-like effect of two dopamine reuptake inhibitors, bupropion and nomifensine, in the forced swimming test (FST) (Yamada et al., 2004; Melo et al., 2011) and a dopamine D2 receptor stimulant was reported to produce antidepressant-like effects in this test (Borsini et al., 1989).

Since the 1990, research has uncovered multiple limitations of the monoamine hypothesis, and its inadequacy has been criticized within the psychiatric community (Hirschfeld, 2000). The antidepressants acutely enhance monoamine availability although this treatment takes weeks to improve the depressive symptoms (Castrén, 2005). Besides, depleting monoamines in healthy patients do not produce mood alteration (Delgado, 2000; Delgado and Moreno, 2000). Recent neurobiological findings indicated that the monoamine hypothesis of depression is limited and new molecular targets are being postulated in the pathophysiology and treatment of depression (Ruhe et al., 2007; Kern et al., 2012).

Glutamatergic System

Besides the well-reported involvement of the monoaminergic system in the pathophysiology and treatment of depression, a growing amount of evidence has pointed that the glutamatergic system plays a key role in mood regulation. The majority of synapses of neocortex, the structure that corresponds to approximately 85% of the brain, belong to the glutamatergic system, suggesting that glutamatergic synapses exceed those of all other neurotransmitters (Douglas and Martin, 2007).

Trullas and Skolnick (1990) published the first study that indicates that the antagonism of N-methyl-D-aspartate (NMDA) receptors, a class of ionotropic glutamate-gated ion channels, produces an antidepressant-like effect. Since then, there have been several preclinical and clinical studies reporting the efficacy of NMDA receptor antagonists of either stand-alone or as an adjunct therapy in depression and depression-related diseases (Szewczyk et al., 2012).

Recent clinical studies have demonstrated that a single subanesthetic dose of ketamine, an ionotropic glutamatergic NMDA receptor antagonist, produces rapid and long-lasting antidepressant response in patients with severe depressive disorder (Berman et al., 2000; Zarate et al., 2006). The acute antidepressant effect of ketamine is of clinical relevance, since all available antidepressants exert their mood-elevating effects only after prolonged administration (several weeks) (Kavalali and Monteggia, 2012).

Several data from animal models have shown that different types of environmental stress/glucocorticoid exposure cause glutamatergic excitotoxicity characterized by an enhanced glutamate release/transmission in limbic/cortical areas, reduced glutamate/glutamine cycling and glial cell density and by dendritic remodeling, reduction of synapses and possibly volumetric reductions resembling those observed in depressed patients (Gorman and Docherty, 2010; Musazzi et al., 2011; Popoli et al., 2011; Sanacora et al., 2012). Magnetic resonance imaging and *postmortem* studies in depressed patients support these findings, reinforcing the notion that the glutamatergic system plays a key role in depression (Kugaya and Sanacora, 2005; Sanacora et al., 2012).

Consistent with the notion that glutamatergic excitotoxicity is implicated in depressive disorders, antiglutamatergic agents, such as riluzole and lamotrigine, have demonstrated antidepressant efficacy (Kugaya and Sanacora, 2005). Moreover, as previously mentioned, the NMDA receptor antagonist ketamine has been used to afford a rapid, efficient and relatively sustained antidepressant effect in treatment-resistant depressive patients (Krystal et al., 2013).

Nitric oxide (NO) is synthesized from L-arginine by a family of isoformic enzymes (eNOS, nNOS and iNOS) known as NO synthase (NOS). In the brain, when NMDA receptors are activated, influx of calcium ion occurs in the neuron, leading to the activation of the enzyme nNOS that converts L-arginine into NO and citrulline. Glutamate neurotoxicity is mediated, in part, by the production of NO via NOS isoforms and mitochondrial damage (Dawson et al., 1991).

Interestingly, several studies have demonstrated that NOS inhibitors exert antidepressant-like effects in animal models (da Silva et al., 2000; Harkin et al., 2003; Joca and Guimarães, 2006), but an antidepressant-like action was observed with L-arginine administration, a substrate for NOS, at low doses, an effect reversed by NOS inhibition, suggesting that NO may play a dual role in the modulation of depression (da Silva et al., 2000; Inan et al., 2004). The plasma NOx levels were reported to be lower in the depressed individuals as compared to healthy subjects and such alteration was abolished by the treatment of the patients with milnacipran, a serotonin and noradrenaline reuptake inhibitor (Ikenouchi-Sugita et al., 2009) and paroxetine, a serotonin reuptake inhibitor (Chrapko et al., 2006). Conversely, plasma nitrate was also shown to be higher in depressed patients than in controls (Suzuki et al., 2001).

The NO modulatory activity of various antidepressants has been demonstrated (Dhir and Kulkarni, 2011). It was shown that the antidepressant responses of mice to several antidepressants, such as imipramine, venlafaxine, escitalopram and duloxetine are dependent, at least in part, on the inhibition, of NO synthesis (Zomkowski et al., 2010, 2012; Krass et al., 2011).

Altogether, the mentioned studies support the hypothesis that the modulation of glutamatergic system and L-arginine-NO pathway plays a significant role in the pathophysiology and treatment of depression.

Oxidative Stress

Oxidative stress is characterized by disequilibrium between oxidant generation and the antioxidant response, resulting in damage to the cells by the excessive reactive oxygen species (ROS) formation (Halliwell, 2006). The brain is particularly vulnerable to the effects of ROS due to its high demand for oxygen, and its abundance of highly peroxidisable substrates associated with a modest antioxidant defense (Halliwell, 2006; Gandhi and Abramov, 2012).

The endogenous detoxification system involves the activity of several antioxidant enzymes. Superoxide dismutase, the first line of defense against ROS, catalyzes the dismutation of superoxide anion radical into hydrogen peroxide (Singh, 1982). It, in turn, can be reduced to water and molecular oxygen by either catalase (Del Río et al., 1992) or glutathione peroxidase (GPx) (Brigelius-Flohé, 1999). The action of GPx converts GSH to glutathione disulfide (GSSG) (Flohé, 1971). The GSSG produced by the GPx-mediated reaction can be restored to GSH through a reaction catalyzed by glutathione reductase (GR), at the expense of the reducing equivalents from NADPH (Kirsch and Groot, 2001).

The glutathione (GSH) system is an important tool mediating protection against several (pro)-oxidant molecules in the brain (Dringen and Hirrlinger, 2003) and alterations is this system seems to be involved in mood regulation. A study by Pal and Dandiya (1994) showed that inescapable foot shock stress in mice elicited a decrease on GSH levels in the cerebral cortex, an effect reversed by antidepressant treatments.

A recent study by Rosa et al. (2013) indicated that the central administration of GSH elicits an antidepressant-like response in the FST and TST in mice.

Indeed, a great amount of evidence has indicated that depressive disorders are accompanied by oxidative stress (Sarandol et al., 2007; Ng et al., 2008, Maes et al., 2011; Behr et al., 2012). Corroborating this notion, conventional and putative antidepressant agents have been reported to exhibit antioxidant properties, reducing oxidative stress associated with depression in preclinical and clinical studies (Zafir and Banu, 2007; Zafir et al., 2009; Behr et al., 2012; Aboul-Fotouh, 2013; Budni et al., 2013; Moretti et al., 2013). These studies suggest that augmentation of antioxidant defenses may be one of the mechanisms underlying the neuroprotective effects of antidepressants.

Inflammation

In the past 20 years numerous studies and meta-analysis emerged indicating a clear association between activation of the immune system, pro-inflammatory cytokines levels, and psychiatric symptoms (Mikova et al., 2001; Tuglu et al., 2003; Dowlati et al., 2010). The relationship between depression and inflammation was established based on several studies which demonstrated that depressive patients have increased inflammatory markers such as interleukin (IL)-6 and tumor necrosis factor-α (TNF-α) (Howren et al., 2009; Dowlati et al., 2010).

Besides, the presence of some inflammatory diseases, such as inflammatory bowel disease, multiple sclerosis, psoriasis, rheumatoid arthritis, and neuro-inflammatory disorders, is reported to increase the risk for the development of depression (Graff et al., 2009). Corroborating the inflammatory hypothesis of depression, it was demonstrated that patients with cancer or chronic hepatitis C, who are treated with inflammatory cytokines, have a higher risk factor for depression (Capuron et al., 2000; Bonaccorso et al., 2001). Moreover, it has been shown that the administration of proinflammatory cytokines, such as interferon (IFN)-α and TNF-α induces a depressive-like behavior in mice (Kaster et al., 2012; Ping et al., 2012).

Nuclear factor erythroid 2-related factor 2 (Nrf2) is a transcription factor that plays a central role in cellular defense against oxidative insults and is reported to play a crucial role in regulating inflammation. Nrf2 knockout mice were shown to exhibit a depressive-like behavior (Martín-de-Saavedra et al., 2013; Muramatsu et al., 2013). Induction of Nrf2 by sulforaphane, in an inflammatory model of depression elicited by lipopolysaccharide afforded antidepressant-like effects (Martín-de-Saavedra et al., 2013). Furthermore, treatment of Nrf2 knockout mice with the anti-inflammatory drug rofecoxib reversed their depressive-like behavior, suggesting that chronic inflammation due to a deletion of Nrf2 can lead to a depressive-like phenotype while induction of Nrf2 could become a new and interesting target to develop novel antidepressant drugs (Martín-de-Saavedra et al., 2013).

Metabolic System

Several studies indicate brain metabolism impairment as a mechanism underlying depression (Mayberg et al., 1994; Moretti et al., 2003). A decrease in the energy metabolism of the frontal lobes and basal ganglia in depressed patients have been reported (Kennedy et al., 2001; Drevets et al., 2002). Depressive patients show a significant decrease on mitochondrial adenosine triphosphate (ATP) production rates in muscle and brain compared to control group (Moore, 1997; Volz et al., 1998; Gardner et al., 2003a,b).

Glucose metabolism in the prefrontal cortex has been negatively correlated with depression severity and poor response to antidepressant therapy (Baxter et al., 1989; Saxena et al., 2003). Furthermore, a study demonstrated metabolic dysfunction (up or downregulation in proteins) in rats submitted to chronic unpredictable mild stress (a depression model): glycolytic proteins (hexokinase↓, glyceraldehyde-3-phosphate dehydrogenase ↑, α-enolase↑↓), tricarboxylic acid cycle proteins (the dihydro-lipoyltransacetylase subunit of pyruvate dehydrogenase↓, 2-oxoglutarate dehydrogenase↓), and two electron transport chain proteins (the NADH dehydrogenase [ubiquinone] iron-sulfur protein 3 subunit of complex I↑, the cytochrome C1 subunit of complex III↑↓) (Yang et al., 2013). Notably, as hexokinase is the initial, non-reversible step of glycolysis, hexokinase downregulation in the prefrontal cortex of chronic unpredictable mild stress rats is consistent with previous positron emission tomography studies showing decreased glucose metabolism in the prefrontal cortex of depressed humans (Subramanian and Miller, 2000). Moreover, a study by Madrigal et al. (2001) reported that chronic stress (immobilization for six hours during 21 days) inhibited complexes I-III and II-III of mitochondrial respiratory chain in rat brain. The inhibition of mitochondrial complexes I, III and IV after chronic unpredicted stress for 40 days was also reported (Rezin et al., 2008). Moreover, this study reported that the complexes I, III and IV were inhibited in cerebral cortex and cerebellum of rats after chronic stress for 21 days. All these data support a metabolic dysfunction on pathophysiology of depression.

The antidepressants imipramine, amitriptyline, doxepine and mianserin increased ATP level in astrocytes culture (Trzeciak et al., 1995). Lithium increased the activity of complexes I-III and II-III of mitochondrial respiratory chain (Lambert et al., 1999; Maurer, Schippel and Volz, 2009). Furthermore, in the hippocampus, cerebellum and striatum of rats administered chronically with bupropion, the activity of complex II was increased (Ferreira et al., 2012). The SSRI citalopram increased the activity of citrate synthase, an enzyme of tricarboxylic acid cycle (Hroudova and Fisar, 2010). Also, chronic administration of paroxetine increased citrate synthase activity in the prefrontal cortex, hippocampus, striatum and cerebral cortex of adult rats (Scaini et al., 2010). Chronic administration of venlafaxine increased succinate dehydrogenase activity in prefrontal cortex of rats (Scaini et al., 2010). Moreover, electroconvulsive shock (an animal model for electroconvulsive therapy) caused an increase on the activities of mitochondrial respiratory chain complexes II and IV (Burigo et al., 2006). It may be concluded that antidepressant treatments have a significant effect on energy metabolism.

Role of Creatine on the Pathophysiology and Treatment of Depression

Creatine is a guanidine-like compound that is synthetized in kidneys, liver and pancreas and, in less quantity, in the brain. Peripherally, the first stage of its synthesis occurs in the kidneys, where the aminoacids glycine and L-arginine undergo a reaction catalyzed by the enzyme L-arginine glycine amidine transferase (AGAT) that results in ornithine and guanidineacetate. Guanidineacetate is transported to liver where occurs the second stage of creatine synthesis, catalyzed by the enzyme guanidineacetate methyltransferase (GAMT), in which a methyl group from S-adenosyl-L-methyonine is transferred to guanidineacetate, forming the creatine molecule (Defalco and Davies, 1961;Walker, 1979; Wyss and Kaddurah-Daouk, 2000).

Studies have shown that the mammalian brain can also synthesize creatine (Van Pilsum et al., 1972). It is well reported that AGAT and GAMT are expressed in the brain (Braissant et al., 2001, 2005; Tachikawa et al., 2004, 2007). AGAT can be expressed in many cell types of CNS as neurons, oligodendrocytes and astrocytes (Braissant et al., 2001; Nakashima et al.,

2005). Furthermore, AGAT is expressed in cells that interconnect periphery to CNS, for example, parenchyma, the cerebrospinal fluid, microcapillary endothelial cells, astrocytes that make contact with the blood brain barrier, choroid plexus and ependymal epithelium (Braissant et al., 2001). GAMT is mainly expressed in astrocytes and oligodendrocytes, but also in neurons (Braissant et al., 2001; Tachikawa et al., 2004; Nakashima et al., 2005). Besides endogenous synthesis, creatine can be obtained by diet through milk, meat and fish ingestion (Walker, 1979; Andres et al., 2008).

Creatine enters in the bloodstream by diffusion and may then be captured in a saturable way by body tissues (red and white blood cells, skeletal and cardiac muscle, brain, retina and spermatozoa) against a gradient concentration by a creatine transporter 1, creatine transporter 2, or by amino acid transporters, as taurine transporter (Wyss and Kaddurah-Daouk, 2000; Abplanalp et al., 2013). Creatine is exocyted in the extracellular medium and captured by other cells through creatine transporter 1 dependent on sodium and chloride ions, playing a role in mediating cellular communication (Almeida et al., 2006).

Inside the cell, creatine can be phosphorylated by the ubiquitous mitochondrial creatine kinase (uMtCK) to phosphocreatine with ATP consumption. Phosphocreatine then diffuses into the cytoplasm and can be converted to creatine by another isoform of creatine kinase, present in the cytosol, the cytosolic brain-type CK (BCK), releasing a molecule of ATP. The creatine molecule formed is transported back to the intermembrane space of mitochondria (Eppenberger et al., 1967; Saks et al., 1978; Bessman and Geiger, 1980; Bessman et al., 1980; Wallimann and Eppenberger, 1985). This system plays an important role in cellular energy metabolism by buffering ATP consumption and improving the flux of high-energy phosphoryls around the cell. A strikingly dissociated pattern of expression was found: uMtCK was found to be ubiquitously and exclusively expressed in neuronal populations, whereas BCK was dominantly expressed in astrocytes, with a low and selective expression in neurons (Lowe et al., 2013).

Alterations in brain creatine levels and creatine kinase activity have been reported in experimental animals submitted to stress as well as in depressive patients. It was reported that a single exposure to the forced swim stress, an inescapable stress condition associated with depressive behavior, decreased creatine and phosphocreatine levels in the prefrontal cortex of mice (Kim et al., 2010). Another study also showed that the forced swim stress increased corticosterone levels (a prevalent condition in experimental models of depression) and decreased creatine levels in the cerebellum of rats (Herring et al., 2008).

Table 1. Studies that report changes in creatine/phosphocreatine levels and creatine kinase activity in animals submitted to stress and/or antidepressants

References	Effect
Czéh et al., 2001	Chronic psychosocial stress model in adult male tree shrews significantly decreased the concentrations of creatine and phosphocreatine (15%) in the brain and the proliferation rate of the granule precursor cells in the dentate gyrus (33%).
Sartorius et al., 2003	Levels of creatine in the hippocampus were increased in rats submitted to learned helplessness model and treated with electroconvulsive shocks.
Herring et al., 2008	Forced swim stress increased corticosterone levels and decreased creatine levels in the cerebellum of rats.
Agostinho et al., 2009	The acute administration of fluoxetine in rats inhibited creatine kinase activity in cerebellum, prefrontal cortex, hippocampus and striatum.
Assis et al., 2009	Ketamine and imipramine decreased the immobility time in the FST and increased creatine kinase activity in striatum, cerebral cortex and cerebellum of rats.
Santos et al., 2009	Chronic administration of paroxetine (SSRI) increased creatine kinase activity in the prefrontal cortex, hippocampus and striatum of adult rats.
Kim et al., 2010	A single exposure to the forced swim stress decreased creatine and phosphocreatine levels in the prefrontal cortex of mice.
Knox et al., 2010	Rats subjected to single prolonged stress showed reduced creatine concentrations in the medial prefrontal cortex as compared to non-stress rats.
Lugenbiel et al., 2010	Rats bred for learned helplessness presented a decrease in the creatine transporter levels in the hippocampus and the treatment with escitalopram reversed this alteration. Conversely, in the prefrontal cortex no alterations in the creatine transporter levels was found, but escitalopram or electroconvulsive shocks increased the creatine transporter levels in this region.
Abelaira et al., 2011	Chronic administration of imipramine and lamotrigine decreased the immobility time in the FST and increased creatine kinase activity in the hippocampus of rats.
Della et al., 2012	The acute treatment with tianeptine increased creatine kinase activity in the prefrontal cortex, but chronic treatment with this antidepressant increased the activity of this enzyme in the hippocampus of rats.
Réus et al., 2012a	The acute and chronic treatments with harmine or imipramine caused an increase on creatine kinase activity in the prefrontal cortex and striatum of rats.
Réus et al., 2012b	The acute and chronic administration of memantine, a compound with antidepressant effect, increased creatine kinase activity in the prefrontal cortex and hippocampus of rats.

An *ex vivo* study using magnetic resonance imaging found that rats subjected to single prolonged stress showed reduced creatine concentrations in the medial prefrontal cortex as compared to non-stress controls (Knox et al., 2010). Studies also have found that subordinate animals exposed repeatedly to psychosocial defeat by dominant animals exhibited significantly less total creatine (creatine + phosphocreatine), reduced hippocampal volume, and impaired neurogenesis (Czéh et al., 2001; Fuchs and Flügge, 2002). Moreover, a study by Sartorius et al. (2003) demonstrated that electroconvulsive shock treatment caused an increase in the hippocampal creatine levels in rats. These results are summarized in Table 1.

Lugenbiel et al. (2010) using a model of congenitally learned helplessness, showed a decline in the creatine transporter levels in the hippocampus of rats and the treatment with escitalopram was able to reverse this alteration. Moreover, congenitally learned helplessness rats treated with escitalopram or electroconvulsive shocks increased the creatine transporter levels in the prefrontal cortex.

Regarding creatine kinase activity, it was reported that acute administration of ketamine, imipramine or harmine, compounds with antidepressant-like action in the forced swim test, increased the activity of the enzyme in the striatum and cerebral cortex of rats (Assis et al., 2009; Réus et al., 2012a). Furthermore, chronic administration of paroxetine (SSRI) increased creatine kinase activity in the prefrontal cortex, hippocampus and striatum of adult rats (Santos et al., 2009). In line with this, a study by Abelaira et al. (2011) showed that creatine kinase activity increased in the amygdala after acute treatment with imipramine in rats. Chronic treatment with imipramine and lamotrigine increased the activity of this enzyme in the hippocampus. Another study also demonstrated that the acute and chronic administration of memantine, a compound with antidepressant properties, increased creatine kinase activity in the prefrontal cortex and hippocampus of rats (Réus et al., 2012b). On the other hand, the acute administration of fluoxetine in rats inhibited creatine kinase activity in cerebellum, prefrontal cortex, hippocampus, striatum and cerebral cortex (Agostinho et al., 2009).

Several clinical studies have also demonstrated changes in creatine/phosphocreatine levels in the brain of depressed patients (Table 2), although other studies demonstrated no changes in the creatine levels (Volz et al., 1998; Auer et al., 2000; Rosenberg et al., 2000; Farchione et al., 2002; Kumar et al., 2002; Michael et al., 2003; Pfleiderer et al., 2003; Sanacora et al., 2004; Brambilla et al., 2005; Ende et al., 2007; Merkl et al., 2011). Some studies demonstrated that creatine levels were increased in the frontal lobe, in the caudate, putamen and thalamus of depressive patients (Gruber et al., 2003; Gabbay et al., 2007). On the other hand, decreased levels of creatine in the anterior cingulate cortex and medial prefrontal cortex of depressive patient were also demonstrated (Mirza et al., 2006; Ventkatramar et al., 2009). In addition, another study reported that phosphocreatine+creatine levels were decreased in the dorsolateral prefrontal cortex in male depressive patients, but were increased in female depressive patients as compared to control individuals (Nery et al., 2009). Also, brain phosphocreatine was significantly decreased in severely (as opposed to mildly) depressive patients (Kato et al., 1992).

Furthermore, a study showed no changes in the creatine levels in the dorsolateral prefrontal cortex of major depressive patients, but a negative correlation was found between the severity of depression and creatine levels (Michael et al., 2003). Dager et al. (2004) also showed an inverse correlation between severity of depression and white matter creatine levels in depressed or mixed-state bipolar patients using two-dimensional proton echo-planar spectroscopic imaging.

Of note, the activity of the enzyme creatine kinase was decreased in the serum of depressed patients (Sora et al., 1986; Segal et al., 2007). However, a study reported a significant increase on the creatine kinase activity in the plasma of depressive, but not euthymic patients (Bălăiță et al., 1990). Another study, demonstrated no significant difference between depressive patients and control group in serum creatine kinase activity, although serum creatine kinase activity was higher in the manic patients than in the depressive patients (Feier et al., 2010) (Table 2). Altogether, these set of preclinical and clinical results suggest that impairment in brain energy metabolism may be implicated in the pathophysiology of depressive disorders and the modulation of energy metabolism by antidepressants could be an important mechanism of action of these drugs.

Table 2. Studies that report changes in creatine/phosphocreatine levels and creatine kinase activity in depressive patients

References	Effect
Sora et al., 1986	Patients with major depression showed lower serum creatine kinase activity.
Bălăiţă et al., 1990	Creatine kinase activity was increased in the plasma of depressive patients.
Kato et al., 1992	Brain phosphocreatine was significantly decreased in severely (as opposed to mildly) depressive patients.
Gruber et al., 2003	Creatine levels were increased in the frontal lobe of depressive patients.
Michael et al., 2003	There were no changes in the creatine levels in the dorsolateral prefrontal cortex in the major depressive patients, but a negative correlation was found between the severity of depression and creatine levels.
Dager et al., 2004	An inverse correlation between severity of depression and white matter creatine levels was reported in depressed or mixed-state bipolar patients using two-dimensional proton echo-planar spectroscopic imaging.
Mirza et al., 2006	Decreases in the creatine levels in the anterior cingulate cortex were detected in depressive patients.
Gabbay et al., 2007	Increases in the creatine levels in the caudate, putamen and thalamus were observed in major depressive patients.
Segal et al., 2007	Unmedicated psychotic major depression patients presented a significantly higher score in Hamilton depression rating scale (HDRS) and lower serum creatine kinase activity compared with unmedicated nonpsychotic major depression.
Ventkatramar et al., 2009	Reduced creatine levels in the medial prefrontal cortex were found in depressive patients as compared to the control group.
Nery et al., 2009	In male depressive patients, phosphocreatine+creatine levels were decreased in the dorsolateral prefrontal cortex, but were increased in female depressive patients as compared to control individuals.
Feier et al., 2010	No significant serum creatine kinase activity difference between depressive patients and control group was found, although serum creatine kinase activity was higher in the manic patients than in the depressive patients.

Antidepressant-Like Effect of Creatine Treatment in Preclinical Studies

The potential role of metabolic impairments in the pathophysiology of depression has motivated researchers to evaluate the efficacy of creatine in preclinical studies and the mechanisms underlying its effects. The antidepressant effect of chronic supplementation with creatine in the FST in rats was firstly shown in a study by Allen et al. (2010). It was also demonstrated that the acute oral administration of creatine elicited an antidepressant-like effect in the TST in mice (Cunha et al., 2012). The antidepressant-like effect of creatine in this test was reported to be dependent on the bioavailability of monoamines in the synaptic cleft, since it was abolished by the pretreatment of mice with α-methyl-p-tyrosine (an inhibitor of tyrosine hydroxylase that reduces dopamine and noradrenaline levels) or p-chlorophenylalanine (an inhibitor of serotonin synthesis) (Cunha et al., 2013b; Cunha et al., 2013c). Moreover, the activation of monoamine receptors could play an important role in the antidepressant response to creatine. In this regard, it was demonstrated that the antidepressant-like effect of creatine in the TST in mice involves a dopaminergic activation (Cunha et al., 2012), besides an activation of α_1-adrenoceptors (Cunha et al., 2013b). Furthermore, antidepressant-like effect of creatine is also likely mediated by an interaction with 5-HT_{1A} receptors, since creatine administered at a subeffective dose in combination with subeffective doses of WAY100635 (a preferential presynaptic 5-HT_{1A} receptor antagonist) or 8-OH-DPAT (a preferential postsynaptic 5-HT_{1A} receptor agonist) produced antidepressant-like effects in the TST (Cunha et al., 2013c).

Preclinical studies also show that the administration of creatine combined with conventional antidepressants such as fluoxetine, paroxetine, citalopram and sertraline (SSRI), amitriptyline (tricyclic antidepressant), imipramine (tricyclic antidepressant), reboxetine (SNRI) produced synergistic effects in the TST, suggesting that an improvement in the response to the antidepressant therapy may occur when creatine is combined with antidepressants (Cunha et al., 2012; Cunha et al., 2013b; Cunha et al., 2013c). In agreement with these studies, it was reported that creatine supplementation potentiated the antidepressant effect of fluoxetine in the FST in rats (Allen et al., 2012). The major studies investigating the antidepressant-like effect of creatine treatment in rodents are presented in Table 3.

Table 3. Summary of findings from preclinical studies in rodents that indicate the antidepressant-like effect of creatine

References	Effect
Allen et al., 2010	Thirty male and 36 female Sprague-Dawley rats were maintained on either chow alone or chow blended with either 2% w/w creatine monohydrate or 4% w/w creatine monohydrate for 5 weeks before the FST. Male rats maintained on 4% creatine displayed increased immobility in the FST, whereas female rats maintained on 4% creatine displayed decreased immobility in the FST.
Allen et al., 2012	Female rats maintained on a 4% creatine diet displayed antidepressant-like effects compared to non-supplemented females. In contrast, creatine did not alter behavior reliably in males. Furthermore, creatine potentiates the sub-acute fluoxetine treatment in rats.
Cunha et al., 2012	Creatine administered by oral route reduced the immobility time in the TST in male and female mice, without affecting locomotor activity. The anti-immobility effect of creatine was prevented by the pretreatment of mice with dopamine D_1 and D_2 receptor antagonists. Subeffective doses of creatine and dopamine D_1 and D_2 receptor agonists or bupropion (a dopamine reuptake inhibitor with subtle activity on noradrenergic reuptake) reduced the immobility time in the TST as compared with either drug alone.
Cunha et al., 2013c	The anti-immobility effect of creatine in the TST was prevented by the pretreatment of mice with an inhibitor of serotonin synthesis. Creatine at a subeffective dose in combination with subeffective doses of WAY100635 (a $5\text{-}HT_{1A}$ receptor antagonist), 8-OH-DPAT (a $5\text{-}HT_{1A}$ receptor agonist) or the SSRI fluoxetine, paroxetine, citalopram and sertraline reduced the immobility time in the TST as compared with either drug alone.
Cunha et al., 2013b	Creatine administered centrally decreased the immobility time in the TST. The anti-immobility effect of peripheral administration of creatine was prevented by the pretreatment of mice with α-methyl-*p*-tyrosine (inhibitor of tyrosine hydroxylase), prazosin (α_1-adrenoceptor antagonist), but not by yohimbine (α_2-adrenoceptor antagonist). Creatine at subeffective dose in combination with subeffective doses of amitriptyline, imipramine, reboxetine or phenylephrine (α_1-adrenoceptor agonist) reduced the immobility time in the TST as compared with either drug alone.

In vivo and *in vitro* studies report that the glutamatergic system is modulated by the treatment with creatine. Of note, creatine modulates the NMDA receptor activation and protects neurons from glutamatergic excitotoxicity (Brewer and Wallimann, 2000; Royes et al., 2008; Genius et al., 2012). Considering that NMDA receptor antagonists are reported to produce antidepressant effects (Berman et al., 2000; Sanacora et al., 2012), the role of NMDA receptors in the antidepressant effect of creatine is under investigation in our laboratory. Interestingly, a study demonstrated that creatine supplementation alone causes a significant decrease on endogenous NO levels (by 33%) and resulted in complete restoration of NO to normal levels in *Drosophila melanogaster* treated with rotenone (Hosamani et al., 2010). On the other hand, an *in vitro* study demonstrated that the incubation with creatine increased the numbers of NOS-immunoreactive neurons (Ducray et al., 2006). Furthermore, a study demonstrated that creatine blocked the increases of NO induced by glutamate and reversed the decreases of NO induced by haloperidol in the neuronal/glial cells (Juravleva et al., 2003, 2005). Then, literature results suggest that creatine may present dual effect on NO synthesis and it could be an important target on the antidepressant effect of creatine supplementation that is also under investigation in our laboratory. As the neuroprotective effect of creatine involves the activation of signaling pathways mediated by multiple intracellular kinases such as PKA, PKC, CaMK-2, PI3K and MEK 1/2 (Cunha et al., 2013a), it remains to be established whether the modulation of these signaling pathways is implicated in the antidepressant-like effect of creatine.

Creatine has an antioxidant effect (Sestili et al., 2011) and can modulate cerebral energy metabolism (Rambo et al., 2013), properties that also may be implicated in its antidepressant effect. In addition, creatine has anti-inflammatory properties, and athletes who are supplemented with creatine decrease the muscle inflammation induced by vigorous muscle training (Santos et al., 2004; Deminice et al., 2013). Since the neuroinflammatory hypothesis of depression is well recognized and established (Mikova et al., 2001; Tuglu et al., 2003; Dowlati et al., 2010), further studies should be conducted to investigate whether the anti-inflammatory property of creatine is implicated in its antidepressant effect.

Antidepressant Effect of Creatine Supplementation in Clinical Studies

After the Olympics in Barcelona (1992) two olympic medalists Linford Christie and Sally Gunnell reported the use of creatine supplement. From that, the use of this compound has become popular among athletes and several studies have shown that creatine supplementation can improve the physical performance of athletes for different sports (Williams et al., 1999). However, only in 2000 some studies began to indicate the therapeutic properties of supplementation with this compound in some diseases (Andres et al., 2008).

Table 4. Summary of major findings from clinical studies that investigate the antidepressant properties of creatine supplementation

References	Effect
Amital et al., 2006	A female patient with post-traumatic stress, depression and fibromyalgia resistant to citalopram showed a decrease in HDRS scores when supplemented with creatine monohydrate (5 grams/day) for 4 weeks.
Roitman et al., 2007	Five patients (2 male and 3 female patients) with treatment-resistant unipolar depression when supplemented with creatine monohydrate (3 g /day in the first week, followed by 5 g/day for another three weeks) showed a significant reduction in HDRS scores.
Kondo et al., 2011	Five female adolescents with unipolar depression resistant to treatment with fluoxetine $\geq$ 8 weeks when supplemented with creatine monohydrate (4 grams/day) for 1-8 weeks significantly reduced Children's Depression Rating Scale-Revised scores.
Lyoo et al., 2012	Fifty-two female patients with unipolar depression treated with the SSRI citalopram and supplemented with creatine monohydrate (3 grams/day) for 2-8 weeks showed a significant reduction in the HDRS scores, as compared to placebo group.
Nemets and Levine, 2013	Fourteen female patients and four male patients with treatment-resistant depression supplemented with creatine monohydrate (5-10 grams/day) in combination with antidepressants for 3 weeks showed no significant reduction in HDRS scores, compared to the placebo group. Despite these negative results, two female patients on creatine augmentation, but none on the placebo, showed early improvement of more than 50% reduction in HDRS scores after 2 weeks of creatine treatment.

Studies demonstrated the antidepressant efficacy of creatine supplementation in depression refractory to treatment with conventional antidepressants (Roitman et al., 2007; Kondo et al., 2011). In former non-medicated patients, the combination of creatine supplementation with SSRIs treatment produced synergistic antidepressant effect as compared to placebo group (Lyoo et al., 2012). Furthermore, creatine supplementation has been shown to improve the symptoms of patients suffering from post-traumatic stress disorder, depression and fibromyalgia (Amital et al., 2006). Also, S-adenosyl-L-methionine, the methyl group donor for the synthesis of creatine, was reported to elicit an antidepressant effect (Papakostas, 2009) and to increase phosphocreatine levels in healthy subjects (Silveri et al., 2003).

A recent study demonstrated that patients with unipolar treatment-resistant, depression supplemented with creatine monohydrate in combination with antidepressants for 3 weeks showed no significant reduction in the severity of depression, as compared to the placebo group, although two female patients with creatine regimen showed early improvement of more than 50% reduction in HDRS scores after 2 weeks of creatine treatment (Nemets and Levine, 2013). A summary of these studies is presented in Table 4.

Conclusion

The past four decades of depression research have focused on the contributions of the monoamines to the pathophysiology and treatment of depression. This chapter presents emerging evidence that the modulation of the metabolic system by creatine may constitute in a novel approach for the management of depressive disorders. However, future research should continue to investigate the molecular mechanisms underlying the antidepressant effect of creatine. This will also contribute to the better understanding of the etiology of major depression.

Acknowledgments

Authors thank Patricia S. Brocardo (Ph.D.) for revising the chapter.

References

Abelaira, H.M., Réus, G.Z., Ribeiro, K.F., Zappellini, G., Ferreira, G.K., Gomes, L.M., Carvalho-Silva, M., Luciano, T.F., Marques, S.O., Streck, E.L., Souza, C.T., Quevedo, J. (2011). Effects of acute and chronic treatment elicited by lamotrigine on behavior, energy metabolism, neurotrophins and signaling cascades in rats. *Neurochem. Int., 59*, 1163-1174.

Aboul-Fotouh, S. (2013). Coenzyme Q10 displays antidepressant-like activity with reduction of hippocampal oxidative/nitrosative DNA damage in chronically stressed rats. *Pharmacol. Biochem. Behav., 104*, 105-112.

Abplanalp, J., Laczko, E., Philp, N.J,, Neidhardt, J., Zuercher, J., Braun, P., Schorderet, D.F., Munier, F.L., Verrey, F., Berger, W., Camargo, S.M., Kloeckener-Gruissem, B. (2013). The cataract and glucosuria associated monocarboxylate transporter MCT12 is a new creatine transporter. *Hum. Mol. Genet., 22*, 3218-3226.

Agostinho, F.R., Scaini, G., Ferreira, G.K., Jeremias, I.C., Réus, G.Z., Rezin, G.T., Castro, A.A., Zugno, A.I., Quevedo, J., Streck, E.L. (2009). Effects of olanzapine, fluoxetine and olanzapine/fluoxetine on creatine kinase activity in rat brain. *Brain Res. Bul., 80*, 337-340.

Albert, P.R. (2012). Transcriptional regulation of the 5-HT$_{1A}$ receptor: implications for mental illness. *Philos. Trans. R. Soc. Lond. B. Biol. Sci., 367*, 2402-2415.

Allen, P. J., D'Anci, K.E., Kanarek, R.B., Renshaw, P.F. (2010). Chronic creatine supplementation alters depression-like behavior in rodents in a sex-dependent manner. *Neuropsychopharmacology, 35*, 534-546.

Allen, P.J., D'Anci, K.E., Kanarek, R.B., Renshaw, P.F. (2012). Sex-specific antidepressant effects of dietary creatine with and without sub-acute fluoxetine in rats. *Pharmacol. Biochem. Behav., 101*, 588-601.

Allen, P.J. (2012). Creatine metabolism and psychiatric disorders: Does creatine supplementation have therapeutic value? *Neurosci. Biobehav. Rev., 36*, 1442-1462.

Almeida, L.S. et al. (2006). Exocytotic release of creatine in rat brain. *Synapse, 60,* 118-123.

Amital, D., Vishne, T., Rubinow, A., Levine, J. (2006). Observed effects of creatine monohydrate in a patient with depression and fibromyalgia. *Am. J. Psychiatry, 163*, 1840-1841.

Andres, R. H., Ducray, A.D., Schlattner, U., Wallimann, T., Widmer, H.R. (2008). Functions and effects of creatine in the central nervous system. *Brain Res. Bull, 76*, 329-343.

Artigas, F., Perez, V., Alvarez, E. (1994). Pindolol induces a rapid improvement of depressed patients treated with serotonin reuptake inhibitors. *Arch. Gen. Psychiatr, 51*, 248-251.

Artigas, F., Romero, L., de Montigny, C., Blier, P. (1996). Acceleration of the effect of selected antidepressant drugs in major depression by 5-HT$_{1A}$ antagonists. *Trends Neurosci.*, *19*, 378-383.

Assis, L. C., Rezin, G.T., Comim, C.M., Valvassori, S.S., Jeremias, I.C., Zugno, A.I., Quevedo, J., Streck, E.L. (2009). Effect of acute administration of ketamine and imipramine on creatine kinase activity in the brain of rats. *Rev. Bras. Psiquiatr*, *31*, 247-252.

Auer, D.P., Pütz, B., Kraft, E., Lipinski, B., Schill, J., Holsboer, F. (2000). Reduced glutamate in the anterior cingulate cortex in depression: an in vivo proton magnetic resonance spectroscopy study. *Biol. Psychiatry*, *47*, 305-13.

Bălăiţă, C., Christodorescu, D., Năstase, R., Iscrulescu, C., Dimian, G. (1990). The serum creatine-kinase as a biologic marker in major depression. *Rom. J. Neurol. Psychiatry*, *28*, 127-34.

Baxter, L. R., Schwartz, J.M., Phelps, M.E., Mazziotta, J.C., Guze, B.H., Selin, C.E., Gerner, R.H., Sumida, R.M. (1989). Reduction of prefrontal cortex glucose metabolism common to three types of depression. *Arch. Gen. Psychiatry*, *46*, 243-250.

Behr, G.A., Moreira, J.C., Frey, B.N. Preclinical and clinical evidence of antioxidant effects of antidepressant agents: implications for the pathophysiology of major depressive disorder. *Oxid. Med. Cell. Longev.*, *2012*, 609421.

Berman, R.M., Moreira, J.C., Frey, B.N. (2000). Antidepressant effects of ketamine in depressed patients. *Biol. Psychiatry, 47*, 254-254.

Berton, O., Nestler, E.J. (2006). New approaches to antidepressant drug discovery: beyond monoamines. *Nat. Rev. Neurosci, 7*, 137-5.

Bessman, S. P., Geiger, P. J. (1980). Compartmentation of hexokinase and creatine phosphokinase, cellular regulation, and insulin action. *Curr. Top Cell. Regul., 16*, 55-86.

Bessman, S.P., Yang, W.C., Geiger, P.J., Erickson-Viitanen, S. (1980). Intimate coupling of creatine phosphokinase and myofibrillar adenosinetriphosphatase. *Biochem. Biophys. Res. Commun, 96*, 1414-1420.

Binfaré, R.W., Mantovani, M., Budni, J., Santos, A.R., Rodrigues, A.L.S. (2010). Involvement of dopamine receptors in the antidepressant-like effect of melatonin in the tail suspension test. *Eur. J. Pharmacol., 638*, 78-83.

Bonaccorso, S., Puzella, A., Marino, V., Pasquini, M., Biondi, M., Artini, M., Almerighi, C., Levrero, M., Egyed, B., Bosmans, E., Meltzer, H.Y., Maes,

M. (2001). Immunotherapy with interferon-alpha in patients affected by chronic hepatitis C induces an intercorrelated stimulation of the cytokine network and an increase in depressive and anxiety symptoms. *Psychiatry Res, 105*, 45-55.

Borsini, F., Lecci, A., Sessarego, A., Frassine, R., Meli, A. (1989). Discovery of antidepressant activity by forced swimming test may depend on pre-exposure of rats to a stressful situation. *Psychopharmacology (Berl), 97*, 183-188.

Braissant, O., Henry, H., Loup, M., Eilers, B., Bachmann, C. (2001). Endogenous synthesis and transport of creatine in the rat brain: an in situ hybridization study. *Brain Res. Mol. Brain Res., 86*, 193-201.

Braissant, O., Henry, H., Villard, A.M., Speer, O., Wallimann, T., Bachmann, C. (2005). Creatine synthesis and transport during rat embryogenesis: spatiotemporal expression of AGAT, GAMT and CT1. *BMC Dev. Biol., 5*, 9.

Brambilla, P., Stanley, J.A., Nicoletti, M.A., Sassi, R.B., Mallinger, A.G., Frank, E., Kupfer, D.J., Keshavan, M.S., Soares, J.C. (2005). 1H Magnetic resonance spectroscopy study of dorsolateral prefrontal cortex in unipolar mood disorder patients. *Psychiatry Res, 138*, 131-139.

Brewer, G. J.; Wallimann, T. W. (2000). Protective effect of the energy precursor creatine against toxicity of glutamate and beta-amyloid in rat hippocampal neurons. *J. Neurochem., 74*, 1968-78.

Brigelius-Flohé, R. (1999). Tissue-specific functions of individual glutathione peroxidases. *Free Radic. Biol. Med., 27*, 951-65.

Brocardo, P.S., Budni, J., Kaster, M.P., Santos, A.R., Rodrigues, A.L.S. (2008). Folic acid administration produces an antidepressant-like effect in mice: evidence for the involvement of the serotonergic and noradrenergic systems. *Neuropharmacology, 54*, 464-473.

Budni, J., Zomkowski, A.D., Engel, D., Santos, D.B., dos Santos, A.A., Moretti, M., Valvassori, S.S., Ornell, F., Quevedo, J., Farina, M., Rodrigues, A.L.S. (2013). Folic acid prevents depressive-like behavior and hippocampal antioxidant imbalance induced by restraint stress in mice. *Exp. Neurol., 240*, 112-121.

Burigo, M., Roza, C.A., Bassani, C., Fagundes, D.A., Rezin, G.T., Feier, G., Dal-Pizzol, F., Quevedo, J., Streck, E.L. (2006). Effect of electroconvulsive shock on mitochondrial respiratory chain in rat brain. *Neurochem. Res., 31*, 1375-1379.

Capuron, L., Ravaud, A., Dantzer, R. (2000). Early depressive symptoms in cancer patients receiving interleukin 2 and/or interferon alfa-2b therapy. *J. Clin Oncol, 18*, 2143-2151.

Caspi, A., Hariri, A.R., Holmes, A., Uher, R., Moffitt, T.E. (2010). Genetic sensitivity to the environment: the case of the serotonin transporter gene and its implications for studying complex diseases and traits. *Am. J. Psychiatry, 167*, 509-527.

Castrén, E. (2005). Is mood chemistry? *Nat Rev Neurosci, 6*, 241-246.

Chrapko, W. Jurasz, P., Radomski, M.W., Archer, S.L., Newman, S.C., Baker, G., Lara, N., Le Mellédo, J.M. (2006). Alteration of decreased plasma NO metabolites and platelet NO synthase activity by paroxetine in depressed patients. *Neuropsychopharmacology, 31,* 1286-1293.

Covington, H.E. 3rd, Vialou, V., Nestler, E.J. (2010). From synapse to nucleus: novel targets for treating depression. *Neuropharmacology, 58,* 683-693.

Cunha, M. P., Machado, D.G., Capra, J.C., Jacinto, J., Bettio, L.E., Rodrigues, A.L.S. (2012). Antidepressant-like effect of creatine in mice involves dopaminergic activation. *J. Psychopharmacol., 26,* 1489-1501.

Cunha, M. P., Martín-de-Saavedra, M.D., Romero, A., Parada, E., Egea, J., Del Barrio, L., Rodrigues, A.L.S., López, M.G. (2013a). Protective effect of creatine against 6-hydroxydopamine-induced cell death in human neuroblastoma SH-SY5Y cells: Involvement of intracellular signaling pathways. *Neuroscience, 238,* 185-194.

Cunha, M. P., Pazini, F.L., Oliveira, Á., Bettio, L.E., Rosa, J.M., Machado, D.G., Rodrigues, A.L.S. (2013b). The activation of alpha1-adrenoceptors is implicated in the antidepressant-like effect of creatine in the tail suspension test. *Prog. Neuropsychopharmacol. Biol. Psychiatry, 44,* 39-50.

Cunha, M. P., Pazini, F.L., Oliveira, Á., Machado, D.G., Rodrigues, A.L.S. (2013c). Evidence for the involvement of 5-HT$_{1A}$ receptor in the acute antidepressant-like effect of creatine in mice. *Brain Res. Bull, 95,* 61-69.

Czéh, B., Michaelis, T., Watanabe, T., Frahm, J., de Biurrun, G., van Kampen, M., Bartolomucci, A., Fuchs, E. (2001). Stress-induced changes in cerebral metabolites, hippocampal volume, and cell proliferation are prevented by antidepressant treatment with tianeptine. *Proc. Natl. Acad. Sci. U S A. 98,* 12796-12801.

da Silva, G.D., Matteussi, A.S., dos Santos, A.R., Calixto, J.B., Rodrigues, A.L.S. (2000). Evidence for dual effects of nitric oxide in the forced swimming test and in the tail suspension test in mice. *Neuroreport* , *11*, 3699-3702.

Dager, S.R., Friedman, S.D., Parow, A., Demopulos, C., Stoll, A.L., Lyoo, I.K., Dunner, D.L., Renshaw, P.F. (2004). Brain metabolic alterations in medication-free patients with bipolar disorder. *Arch. Gen. Psychiatry*, *61*, 450-458.

Dawson, V.L., Dawson, T.M., London, E.D., Bredt, D.S., Snyder, S.H. (1991). Nitric oxide mediates glutamate neurotoxicity in primary cortical cultures. *Proc. Natl. Acad. Sci. U S A*, *88*, 6368-6371.

Defalco, A.J., Davies, R.K. (1961). The synthesis of creatine by the brain of the intact rat. *J. Neurochem.*, *7*, 308-312.

del Río, L.A., Sandalio, L.M., Palma, J.M., Bueno, P., Corpas, F.J. (1992). Metabolism of oxygen radicals in peroxisomes and cellular implications. *Free Radic. Biol. Med.*, *13*, 557-580.

Delgado, P.L., Moreno, F.A. (2000). Role of norepinephrine in depression. *J. Clin. Psychiatry*, *61*, 5-12.

Delgado, P.L. (2000). Depression: the case for a monoamine deficiency. *J. Clin. Psychiatry*, *61*, 7-11.

Della, F.P., Abelaira, H.M., Réus, G.Z., Ribeiro, K.F., Antunes, A.R., Scaini, G., Jeremias, I.C., dos Santos, L.M., Jeremias, G.C., Streck, E.L., Quevedo, J. (2012). Tianeptine treatment induces antidepressive-like effects and alters BDNF and energy metabolism in the brain of rats. *Behav. Brain Res.*, *233*, 526-535.

Deminice, R., Rosa, F.T., Franco, G.S., Jordao, A.A., de Freitas, E.C. (2013). Effects of creatine supplementation on oxidative stress and inflammatory markers after repeated-sprint exercise in humans. *Nutrition.*,*29*,1127-1132.

Deupree, J.D., Reed, A.L., Bylund, D.B. (2007). Differential effects of the tricyclic antidepressant desipramine on the density of adrenergic receptors in juvenile and adult rats. *J. Pharmacol. Exp. Ther.*, *321*, 770–776.

Dhir, A., Kulkarni, S.K. (2011). Nitric oxide and major depression. *Nitric. Oxide*, *24*, 125-131.

Dick, D.M., Riley, B., Kendler, K.S. (2010). Nature and nurture in neuropsychiatric genetics: where do we stand? *Dialogues Clin. Neurosci.*,*12*, 7-23.

Douglas, R.J., Martin, K.A. (2007). Mapping the matrix: the ways of neocortex. *Neuron, 56*, 226-238.

Dowlati, Y., Herrmann, N., Swardfager, W., Liu, H., Sham, L., Reim, E.K., Lanctôt, K.L. (2010). A meta-analysis of cytokines in major depression. *Biol. Psychiatry*, *67*, 446-457.

Drevets, W. C., Price, J.L., Bardgett, M.E., Reich, T., Todd, R.D., Raichle, M.E. (2002). Glucose metabolism in the amygdala in depression: relationship to diagnostic subtype and plasma cortisol levels. *Pharmacol. Biochem Behav*, *71*, 431-447.

Dringen, R., Hirrlinger, J. (2003). Glutathione pathways in the brain. *Biol. Chem.*, *384*, 505-516.

Ducray, A., Kipfer, S., Huber, A.W., Andres, R.H., Seiler, R.W., Schlattner, U., Wallimann, T., Widmer, H.R. (2006). Creatine and neurotrophin-4/5 promote survival of nitric oxide synthase-expressing interneurons in striatal cultures. *Neurosci. Lett*, *395*, 57-62.

Elhwuegi, A.S. (2004). Central monoamines and their role in major depression. *Prog Neuropsychopharmacol Biol Psychiatry*, *28*, 435-451.

Ende, G., Demirakca, T., Walter, S., Wokrina, T., Sartorius, A., Wildgruber, D., Henn, F.A. (2007). Subcortical and medial temporal MR-detectable metabolite abnormalities in unipolar major depression. *Eur. Arch. Psychiatry Clin. Neurosci. 257*, 36–39.

Eppenberger, M. E., Eppenberger, H.M., Kaplan, N.O. (1967). Evolution of creatine kinase. *Nature*, *214*, 239-241.

Farchione, T.R., Moore, G.J., Rosenberg, D.R. (2002) Proton magnetic resonance spectroscopic imaging in pediatric major depression. *Biol. Psychiatry*, *52*, 86-92.

Fava, M., Kendler, K.S. (2000). Major depressive disorder. *Neuron, 28*, 335-341.

Feier, G., Valvassori, S.S., Rezin, G.T., Búrigo, M., Streck, E.L., Kapczinski, F., Quevedo, J. (2010). Creatine kinase levels in patients with bipolar disorder: depressive, manic, and euthymic phases. *Rev. Bras. Psiquiatr*, *33*, 171-175.

Ferreira, G.K., Rezin, G.T., Cardoso, M.R., Gonçalves, C.L., Borges, L.S., Vieira, J.S., Gomes, L.M., Zugno, A.I., Quevedo, J., Streck, E.L. (2012). Brain energy metabolism is increased by chronic administration of bupropion. *Acta Neuropsychiatrica*, *24*, 115-121.

Flohé, L. (1971). Glutathione peroxidase: enzymology and biological aspects. *Klin. Wochenschr.*, *49*, 669-683.

Frazer, A. (1997). Pharmacology of antidepressants. *J Clin Psychopharmacol*, *1*, 2S-18S.

Fuchs, E., Flügge, G. (2002). Psychosocial stress induces molecular and structural alterations in the brain - How animal experiments help to understand pathomechanisms of depressive illnesses. *Z Psychosom. Med. Psychother*, *47*, 80-97.

Gabbay, V., Hess, D.A., Liu, S., Babb, J.S., Klein, R.G., Gonen, O. (2007). Lateralized caudate metabolic abnormalities in adolescent major depressive disorder: a proton MR spectroscopy study. *Am. J. Psychiatry*, *164*, 1881-1889.

Gambarana, C., Ghiglieri, O., Tagliamonte, A., D'Alessandro, N., de Montis M,G. (1995). Crucial role of D1 dopamine receptors in mediating the antidepressant effect of imipramine. *Pharmacol. Biochem. Behav.*, *50*, 147-151.

Gandhi, S., Abramov, A.Y. (2012). Mechanism of oxidative stress in neurodegeneration. *Oxid. Med. Cell. Longev.*, *2012*, 428010.

Gardner, A., Johansson, A., Wibom, R., Nennesmo, I., von Döbeln, U., Hagenfeldt, L., Hällström, T. (2003a). Alterations of mitochondrial function and correlations with personality traits in selected major depressive disorder patients. *J. Affect. Disord*, *76*, 55-68.

Gardner, A., Pagani, M., Wibom, R., Nennesmo, I., Jacobsson, H., Hällström, T. (2003b). Alterations of rCBF and mitochondrial dysfunction in major depressive disorder: a case report. *Acta. Psychiatr. Scand.*, *107*, 233-239.

Genius, J., Geiger, J., Bender, A., Möller, H.J., Klopstock, T., Rujescu, D. (2012). Creatine protects against excitoxicity in an in vitro model of neurodegeneration. *PLoS One*, *7*, e30554.

Gorman, J.M., Docherty, J.P. (2010). A hypothesized role for dendritic remodeling in the etiology of mood and anxiety disorders. *J. Neuropsychiatry Clin. Neurosci.*, *22*, 256-264.

Graff, L.A., Walker, J.R., Bernstein, C.N. (2009). Depression and anxiety in inflammatory bowel disease: a review of comorbidity and management. *Inflamm. Bowel. Dis.*, *15*, 1105-1118.

Gruber, S., Frey, R., Mlynárik, V., Stadlbauer, A., Heiden, A., Kasper, S., Kemp, G.J., Moser, E. (2003). Quantification of metabolic differences in the frontal brain of depressive patients and controls obtained by 1H-MRS at 3 Tesla. *Invest. Radiol.*, *38*, 403-408.

Halliwell, B. (2006). Oxidative stress and neurodegeneration: where are we now? *J. Neurochem.*, *97*, 1634-1658.

Harkin, A.J., Connor, T.J., Walsh, M., St John N., Kelly, J.P. (2003). Serotonergic mediation of the antidepressant-like effects of nitric oxide synthase inhibitors. *Neuropharmacology*, *44*, 616-623.

Hensler, J.G. (2002). Differential regulation of 5-HT$_{1A}$ receptor-G protein interactions in brain following chronic antidepressant administration. *Neuropsychopharmacology, 26*, 565–573.

Herring, N.R., Schaefer, T.L., Tang, P.H., Skelton, M.R., Lucot, J.P., Gudelsky, G.A., Vorhee, C.V., Williams, M.T. (2008). Comparison of time-dependent effects of (+)-methamphetamine or forced swim on monoamines, corticosterone, glucose, creatine, and creatinine in rats. *BMC Neurosci, 30*, 49.

Hirschfeld, R.M. (2000). History and evolution of the monoamine hypothesis of depression. *J Clin Psychiatry, 61*, 4-6.

Holden, C. (2005). Sex and the suffering brain. *Science, 308*, 1574.

Holsboer, F., Barden N. (1996). Antidepressants and hypothalamic–pituitary–adrenocortical regulation. *Endocr. Rev., 17*, 187-205.

Hosamani, R., Ramesh, S.R., Muralidhara. (2010). Attenuation of rotenone-induced mitochondrial oxidative damage and neurotoxicty in Drosophila melanogaster supplemented with creatine. *Neurochem. Res, 35*, 1402-1412.

Howren, M.B., Lamkin, D.M., Suls, J. (2009). Associations of depression with C-reactive protein, IL-1, and IL-6: a meta-analysis. *Psychosom. Med., 71*, 171-186.

Hroudova, J., Fisar, Z. (2010). Activities of respiratory chain complexes and citrate synthase influenced by pharmacologically different antidepressants and mood stabilizers. *Neuro. Endocrinol. Lett., 31*, 336-342.

Ikenouchi-Sugita, A., Yoshimura, R., Hori, H., Umene-Nakano, W., Ueda, N., Nakamura, J. (2009). Effects of antidepressants on plasma metabolites of nitric oxide in major depressive disorder: comparison between milnacipran and paroxetine. *Prog. Neuropsychopharmacol. Biol. Psychiatry, 33*, 1451-1143.

Inan, S.Y., Yalcin, I., Aksu, F. (2004). Dual effects of nitric oxide in the mouse forced swimming test: possible contribution of nitric oxide-mediated serotonin release and potassium channel modulation. *Pharmacol. Biochem. Behav., 77*, 457-464.

Joca, S.R., Guimarães, F.S. (2006). Inhibition of neuronal nitric oxide synthase in the rat hippocampus induces antidepressant-like effects. *Psychopharmacology, 185*, 298-305.

Juravleva, E., Barbakadze, T.,, Mikeladze1, D., Kekelidze, T. (2003). Creatine prevents the cytotoxicity of haloperidol by alteration of NO/Ras/NF-κB system. In: Kekelidze T, Holtzman D, editors. Creatine kinase and brain energy metabolism: function and disease. *NATO Science Series: Life and Behavioral Science*. Amsterdam, IOS Press, *343*, 113-119.

Juravleva, E., Barbakadze, T., Mikeladze, D., Kekelidze, T. (2005). Creatine enhances survival of glutamate-treated neuronal/glial cells, modulates Ras/NF-kappaB signaling, and increases the generation of reactive oxygen species. *J. Neurosci. Res.*, *79*, 224-230.

Kaster, M.P., Rosa, A.O., Santos, A.R., Rodrigues, A.L.S. (2005). Involvement of nitric oxide-cGMP pathway in the antidepressant-like effects of adenosine in the forced swimming test. *Int. J. Neuropsychopharmacol.*, *8*, 601-606.

Kaster, M.P. Gadotti, V.M., Calixto, J.B., Santos, A.R., Rodrigues, A.L.S. (2012). Depressive-like behavior induced by tumor necrosis factor-α in mice. *Neuropharmacology*, *62*, 419-426.

Kato, T., Takahashi, S., Shioiri, T., Inubushi, T. (1992). Brain phosphorous metabolism in depressive disorders detected by phosphorus-31 magnetic resonance spectroscopy. *J. Affect. Disord*, *26*, 223-230.

Kavalali, E.T., Monteggia, L.M. (2012). Synaptic mechanisms underlying rapid antidepressant action of ketamine. *Am. J. Psychiatriy*, 169, 1150-1156.

Kennedy, S. H., Evans, K.R., Krüger, S., Mayberg, H.S., Meyer, J.H., McCann, S., Arifuzzman, A.I., Houle, S., Vaccarino, F.J. (2001). Changes in regional brain glucose metabolism measured with positron emission tomography after paroxetine treatment of major depression. *Am. J. Psychiatry*, *158*, p. 899-905.

Kern, N., Sheldrick, A.J., Schmidt, F.M., Minkwitz, J. (2012). Neurobiology of depression and novel antidepressant drug targets. *Curr. Pharm. Des.*, *18*, 5791-5801.

Kessler, D. Sharp, D., Lewis, G. (2005). Screening for depression in primary care. *Br. J. Gen. Pract.*, *55*, 659-660.

Kim, S.Y., Lee, Y.J., Kim, H., Lee, D.W., Woo, D.C., Choi, C.B., Chae, J.H., Choe, B.Y. (2010). Desipramine attenuates forced swim test-induced behavioral and neurochemical alterations in mice: an in vivo(1)H-MRS study at 9.4T. *Brain Res*, *1348*, 105-113.

Kirsch, M., De Groot, H. (2001). NAD(P)H, a directly operating antioxidant? *FASEB J. 15*, 1569-1574.

Klimek, V., Zak-Knapik, J., Maj, J. (1991). Antidepressants given repeatedly increase the alpha 1-adrenoceptor agonist affinity in the rat brain. *Pol. J. Pharmacol. Pharm.*, *43*, 347-352.

Knox, D., Perrine, S.A., George, S.A., Galloway, M.P., Liberzon, I. (2010). Single prolonged stress decreases glutamate, glutamine, and creatine concentrations in the rat medial prefrontal cortex. *Neurosci. Lett., 480*, 16-20.

Kondo, D.G., Sung, Y.H., Hellem, T.L., Fiedler, K.K., Shi, X., Jeong, E.K., Renshaw, P.F. (2011). Open-label adjunctive creatine for female adolescents with SSRI-resistant major depressive disorder: a 31-phosphorus magnetic resonance spectroscopy study. *J. Affect. Disord., 135*, 354-361.

Krass, M. Wegener, G., Vasar, E., Volke, V. (2011). The antidepressant action of imipramine and venlafaxine involves suppression of nitric oxide synthesis. *Behav. Brain. Res., 218*, 57-63.

Krishnan, V., Nestler, E.J. (2008). The molecular neurobiology of depression. *Nature, 455*, 894-902.

Krystal, J.H., Sanacora, G., Duman, R.S. (2013). Rapid-acting glutamatergic antidepressants: the path to ketamine and beyond. *Biol. Psychiatry, 73*, 1133-1141.

Kugaya, A., Sanacora, G. (2005). Beyond monoamines: glutamatergic function in mood disorders. *CNS Spectr, 10*, 808-819.

Kumar, A., Thomas, A., Lavretsky, H., Yue, K., Huda, A., Curran, J., Venkatraman, T., Estanol, L., Mintz, J., Mega, M., Toga, A. (2002). Frontal white matter biochemical abnormalities in late-life major depression detected with proton magnetic resonance spectroscopy. *Am. J. Psychiatry 159*, 630–636.

Lambert, P. D., McGirr, K.M., Ely, T.D,. Kilts, C.D, Kuhar, M.J. (1999). Chronic lithium treatment decreases neuronal activity in the nucleus accumbens and cingulate cortex of the rat. *Neuropsychopharmacology, 21*, 229-237.

Lowe, M.T. (2013). Dissociated expression of mitochondrial and cytosolic creatine kinases in the human brain: a new perspective on the role of creatine in brain energy metabolism. *J Cereb Blood Flow Metab, 33*, 1295-1306.

Lugenbiel, P., Sartorius, A., Vollmayr, B., Schloss, P. (2010). Creatine transporter expression after antidepressant therapy in rats bred for learned helplessness. *World J. Biol. Psychiatry, 11*, 329-333.

Lyoo, I. K., Yoon, S., Kim, T.S., Hwang, J., Kim, J.E., Won, W., Bae, S., Renshaw, P.F. (2012). A randomized, double-blind placebo-controlled trial of oral creatine monohydrate augmentation for enhanced response to a selective serotonin reuptake inhibitor in women with major depressive disorder. *Am. J. Psychiatry, 169,* 937-945.

Madrigal, J.L., Olivenza, R., Moro, M.A., Lizasoain, I., Lorenzo, P., Rodrigo, J., Leza, J.C. (2001). Glutathione depletion, lipid peroxidation and mitochondrial dysfunction are induced by chronic stress in rat brain. *Neuropsychopharmacology, 24,* 420-429.

Maes, M., Fisar, Z., Medina, M., Scapagnini, G., Nowak, G., Berk, M. (2012). New drug targets in depression: inflammatory, cell-mediated immune, oxidative and nitrosative stress, mitochondrial, antioxidant, and neuroprogressive pathways. And new drug candidates-Nrf2 activators and GSK-3 inhibitors. *Inflammopharmacology, 20,* 127-150.

Mann, J.J., Malone, K.M., Sweeney, J.A., Brown, R.P., Linnoila, M., Stanley, B., Stanley, M. (1996). Attempted suicide characteristics and cerebrospinal fluid amine metabolites in depressed inpatients. *Neuropsychopharmacology. 15,* 576-586.

Martín-de-Saavedra, M.D., Budni, J., Cunha, M.P., Gómez-Rangel, V., Lorrio, S., Del Barrio, L., Lastres-Becker, I., Parada, E., Tordera, R.M., Rodrigues, A.L., Cuadrado, A., López, M.G. (2013). Nrf2 participates in depressive disorders through an anti-inflammatory mechanism. *Psychoneuroendocrinology.* In Press. doi: 10.1016/j.psyneuen.2013. 03.020.

Masand, P.S. (2003) Tolerability and adherence issues in antidepressant therapy. *Clin. Ther, 25,* 2289-2304.

Maurer, I. C., Schippel, P., Volz, H. P. (2009). Lithium-induced enhancement of mitochondrial oxidative phosphorylation in human brain tissue. *Bipolar. Disord., 11,* 515-522.

Mayberg, H.S., Lewis, P.J., Regenold, W., Wagner, H.N. (1994). Paralimbic hypoperfusion in unipolar depression. *J. Nucl. Med., 35,* 929-934.

McEwen, B.S., Eiland, L., Hunter, R.G., Miller, M,M. (2012). Stress and anxiety: structural plasticity and epigenetic regulation as a consequence of stress. *Neuropharmacology, 62,* 3-12.

Melo, F.H., Moura. B.A., de Sousa, D.P., de Vasconcelos, S.M., Macedo, D.S., Fonteles, M.M., Viana, G.S., de Sousa, F.C. (2011). Antidepressant-like effect of carvacrol (5-Isopropyl-2-methylphenol) in mice: involvement of dopaminergic system. *Fundam Clin. Pharmacol. 25,* 362-367.

Merkl, A., Schubert, F., Quante, A., Luborzewski, A., Brakemeier, E.L., Grimm, S., Heuser, I., Bajbouj, M. (2011). Abnormal cingulate and prefrontal cortical neurochemistry in major depression after electroconvulsive therapy. *Biol. Psychiatry*, *69*, 772-779.

Michael, N., Erfurth, A., Ohrmann, P., Arolt, V., Heindel, W., Pfleiderer, B,. (2003). Metabolic changes within the left dorsolateral prefrontal cortex occurring with electroconvulsive therapy in patients with treatment resistant unipolar depression. *Psychol. Med.*, *33*, 1277-1284.

Mikova, O., Yakimova, R., Bosmans, E., Kenis, G., Maes, M. (2001). Increased serum tumor necrosis factor alpha concentrations in major depression and multiple sclerosis. *Eur. Neuropsychopharmacol.*, *11*, 203-208.

Mirza, Y., O'Neill, J., Smith, E.A., Russell, A., Smith, J.M., Banerjee, S.P., Bhandari, R., Boyd, C., Rose, M., Ivey, J., Renshaw, P.F., Rosenberg, D.R. (2006). Increased medial thalamic creatine-phosphocreatine found by proton magnetic resonance spectroscopy in children with obsessive-compulsive disorder versus major depression and healthy controls. *J. Child Neurol.. 21*, 106-111.

Moore, R.G. (1997). Improving the treatment of depression in primary care: problems and prospects. *Br. J. Gen. Pract.*, *47*, 587-590.

Moretti, A., Gorini, A., Villa, R. F. (2003). Affective disorders, antidepressant drugs and brain metabolism. *Mol. Psychiatry*, *8*, 773-785.

Moretti, M., Budni, J., Dos Santos, D.B., Antunes, A., Daufenbach, J.F., Manosso, L.M., Farina, M., Rodrigues, A.L. (2013). Protective effects of ascorbic acid on behavior and oxidative status of restraint-stressed mice. *J. Mol. Neurosci.*, *49*, 68-79.

Morrison, K. E., Curry, D.W, Coopper, M.A. (2012). Social status alters defeat-induced neural activation in Syrian hamsters. *Neuroscience, 210*, 168-178.

Muramatsu, H., Katsuoka, F., Toide, K., Shimizu, Y., Furusako, S., Yamamoto, M. (2013) Nrf2 deficiency leads to behavioral, neurochemical and transcriptional changes in mice. Genes Cells. In press, doi: 10.1111/gtc.12083.

Musazzi, L., Racagni, G., Popoli, M. (2011). Environmental stress, glucocorticoids and glutamate release: effects of antidepressant drugs. *Neurochem. Int.*, *59*, 138-149.

Nakajima, S., Suzuki, T., Watanabe, K., Kashima, H., Uchida, H. (2010). Accelerating response to antidepressant treatment in depression: a review and clinical suggestions. *Prog. Neuropsychopharmacol. Biol. Psychiatry*, *34*, 259-264.

Nakashima, T., Tomi, M., Tachikawa, M., Watanabe, M., Terasaki, T., Hosoya, K. (2005). Evidence for creatine biosynthesis in Muller glia. *Glia*. v. 52, p. 47-52.

Nemeroff, C.B. (2007) Prevalence and management of treatment-resistant depression. *J. Clin. Psychiatry*, *8*, 17-25.

Nemets, B., Levine, J. (2013). A pilot dose-finding clinical trial of creatine monohydrate augmentation to SSRIs/SNRIs/NASA antidepressant treatment in major depression. *Int. Clin. Psychopharmacol.*, *28*, 127-133.

Nery, F.G., Stanley, J.A., Chen, H.H., Hatch, J.P., Nicoletti, M.A., Monkul, E.S., Matsuo, K., Caetano, S.C., Peluso, M.A., Najt, P., Soares, J.C. (2009). Normal metabolite levels in the left dorsolateral prefrontal cortex of unmedicated major depressive disorder patients: a single voxel (1)H spectroscopy study. *Psychiatry Res. 174*, 177-183.

Nestler, E.J., Barrot, M., DiLeone, R.J., Eisch, A.J., Gold, S.J., Monteggia, L.M. (2002). Neurobiology of depression. *Neuron, 34*, 13-25.

Ng, F., Berk, M., Dean, O., Bush, A.I. (2008) Oxidative stress in psychiatric disorders: evidence base and therapeutic implications. *Int. J. Neuropsychopharmacol.*, *11*, 851-876.

Nowak, G., Przegaliński, E. (1988). Effect of repeated treatment with antidepressant drugs and electroconvulsive shock (ECS) on [3H] prazosin binding to different rat brain structures. *J. Neural. Transm., 71,* 57-64.

Pal, S.N., Dandiya, P.C. (1994). Glutathione as a cerebral substrate in depressive behavior. *Pharmacol. Biochem. Behav., 48*, 845-851.

Papakostas, G. I. (2009). Evidence for S-adenosyl-L-methionine (SAM-e) for the treatment of major depressive disorder. *J. Clin. Psychiatry, 70*, 18-22.

Pfleiderer, B., Michael, N., Erfurth, A., Ohrmann, P., Hohmann, U., Wolgast, M., Fiebich, M., Arolt, V., Heindel, W. (2003). Effective electroconvulsive therapy reverses glutamate/glutamine deficit in the left anterior cingulum of unipolar depressed patients. *Psychiatry Res: Neuroimaging, 122,* 185–192.

Ping, F., Shang, J., Zhou, J., Zhang, H., Zhang, L. (2012). 5-HT$_{1A}$ receptor and apoptosis contribute to interferon-α-induced "depressive-like" behavior in mice. *Neurosci. Lett., 514*, 173-178.

Pittenger, C., Duman, R.S. (2008). Stress, depression, and neuroplasticity: a convergence of mechanisms. *Neuropsychopharmacology, 33*, 88-109.

Popoli, M., Yan, Z., McEwen, B.S., Sanacora, G. (2011). The stressed synapse: the impact of stress and glucocorticoids on glutamate transmission. *Nat. Rev. Neurosci., 13*, 22-37.

Rambo, L. M., Ribeiro, L.R., Della-Pace, I.D., Stamm, D.N., da Rosa Gerbatin, R., Prigol, M., Pinton, S., Nogueira, C.W., Furian, A.F,, Oliveira, M.S., Fighera, M.R., Royes, L.F. (2013). Acute creatine administration improves mitochondrial membrane potential and protects against pentylenetetrazol-induced seizures. *Amino Acids*, 44, 857-868.

Rehavi, M., Ramot, O., Yavetz, B., Sokolovsky, M. (1980). Amitriptyline: long-term treatment elevates alpha-adrenergic and muscarinic receptor binding in mouse brain. *Brain Res., 194*, 443–453.

Renard, C.E., Fiocco, A.J., Clenet, F., Hascoet, M., Bourin, M. (2001). Is dopamine implicated in the antidepressant-like effects of selective serotonin reuptake inhibitors in the mouse forced swimming test? *Psychopharmacology (Berl), 159*, 42-50.

Réus, G.Z., Stringari, R.B., Gonçalves, C.L., Scaini, G., Carvalho-Silva, M., Jeremias, G.C., Jeremias, I.C., Ferreira, G.K., Streck, E.L., Hallak, J.E., Zuardi, A.W., Crippa, J.A., Quevedo, J. (2012a). Administration of harmine and imipramine alters creatine kinase and mitochondrial respiratory chain activities in the rat brain. *Depress Res. Treat, 2012*, 987397.

Réus, G.Z., Stringari, R.B., Rezin, G.T., Fraga, D.B., Daufenbach, J.F., Scaini, G., Benedet, J., Rochi, N., Streck, E.L., Quevedo, J. (2012b). Administration of memantine and imipramine alters mitochondrial respiratory chain and creatine kinase activities in rat brain. *J. Neural. Transm, 119*, 481-491.

Rezin, G. T., Cardoso MR, Gonçalves CL, Scaini, G., Fraga, D.B., Riegel, R.E., Comim, C.M., Quevedo, J., Streck, E.L. (2008). Inhibition of mitochondrial respiratory chain in brain of rats subjected to an experimental model of depression. *Neurochem. Int., 53*, 395-400.

Rogóz, Z., Skuza, G., Leśkiewicz, M., Budziszewska, B. (2008). Effects of co-administration of fluoxetine or tianeptine with metyrapone on immobility time and plasma corticosterone concentration in rats subjected to the forced swim test. *Pharmacol. Rep., 60*, 880-888.

Roitman, S., Green, T., Osher, Y., Karni, N., Levine, J. (2007). Creatine monohydrate in resistant depression: a preliminary study. *Bipolar. Disord., 9*, 754-758.

Rosa, J.M., Dafre, A.L., Rodrigues, A.L.S. (2013). Antidepressant-like responses in the forced swimming test elicited by glutathione and redox modulation. *Behav. Brain Res, 253*,165-172.

Rosenberg, D.R., L.D., Paulson, N., Seraji-Bozorgzad, I.B., Wilds, C.M., Stewart, G.J., Moore. (2000). Brain chemistry in pediatric depression. *Biol. Psychiatry.* 47, 95S.

Royes, L. F., Fighera, M.R., Furian, A.F., Oliveira, M.S., Fiorenza, N.G., Ferreira, J., da Silva, A.C., Priel, M.R., Ueda, E.S., Calixto, J.B., Cavalheiro, E.A., Mello, C.F. (2008). Neuromodulatory effect of creatine on extracellular action potentials in rat hippocampus: role of NMDA receptors. *Neurochem. Int.*, 53, 33-37.

Ruhe, H.G., Mason, N.S., Schene, A.H. (2007). Mood is indirectly related to serotonin, norepinephrine and dopamine levels in humans: a meta-analysis of monoamine depletion studies. *Mol. Psychiatry*, *12*, 331–359.

Saks, V. A., Rosenshtraukh, L.V., Smirnov, V.N., Chazov, E.I. (1978). Role of creatine phosphokinase in cellular function and metabolism. *Can. J. Physiol. Pharmacol*, *56*, 691-706.

Sanacora, G., Gueorguieva. R., Epperson, C.N., Wu, Y.T., Appel, M., Rothman, D.L., Krystal, J.H., Mason, G.F. (2004). Subtype-specific alterations of gamma-aminobutyric acid and glutamate in patients with major depression. *Arch. Gen. Psychiatry, 61,* 705-713.

Sanacora, G., Treccani, G., Popoli, M. (2012). Towards a glutamate hypothesis of depression: an emerging frontier of neuropsychopharmacology for mood disorders. *Neuropharmacology, 62,* 63-77.

Santos, R.V., Bassit, R.A., Caperuto, E.C., Rosa, L.F.C. (2004). The effect of creatine supplementation upon inflammatory and muscle soreness markers after a 30km race. *Life Sci, 75,* 1917-1924.

Santos, P. M., Scaini, G., Rezin, G.T., Benedet, J., Rochi, N., Jeremias, G.C., Carvalho-Silva, M., Quevedo, J., Streck, E.L. (2009). Brain creatine kinase activity is increased by chronic administration of paroxetine. *Brain Res. Bull, 80,* 327-330.

Sarandol, A., Sarandol, E., Eker, S.S., Erdinc, S., Vatansever, E., Kirli, S. (2007). Major depressive disorder is accompanied with oxidative stress: short-term antidepressant treatment does not alter oxidative-antioxidative systems. *Hum. Psychopharmacol., 22,* 67-73.

Sartorius, A., Vollmayr, B., Neumann-Haefelin, C., Ende, G., Hoehn, M., Henn, F.A. (2003). Specific creatine rise in learned helplessness induced by electroconvulsive shock treatment. *Neuroreport, 14,* 2199-2201.

Saxena, S., Brody, A.L., Ho, M.L., Zohrabi, N., Maidment, K.M., Baxter, L.R. Jr. (2003). Differential brain metabolic predictors of response to paroxetine in obsessive-compulsive disorder versus major depression. *Am. J. Psychiatry, 160,* 522-532.

Scaini, G., Santos, P.M., Benedet, J., Rochi, N., Gomes, L.M., Borges, L.S., Rezin, G.T., Pezente, D.P., Quevedo, J., Streck, E.L. (2010). Evaluation of Krebs cycle enzymes in the brain of rats after chronic administration of antidepressants. *Brain Res. Bull, 82,* 224-227.

Schildkraut, J.J. (1974). Biogenic amines and affective disorders. *Annu. Rev. Med., 25,* 333-348.

Schmidt, H. D., Duman, R. S. (2007). The role of neurotrophic factors in adult hippocampal neurogenesis, antidepressant treatments and animal models of depressive-like behavior. *Behav. Pharmacol., 18,* 391-418.

Segal, M., Avital, A., Drobot, M., Lukanin, A., Derevenski, A., Sandbank, S., Weizman, A. (2007). Serum creatine kinase level in unmedicated nonpsychotic, psychotic, bipolar and schizoaffective depressed patients. *Eur. Neuropsychopharmacol., 17,* 194-198.

Sestili, P., Martinelli, C., Bravi, G., Piccoli, G., Curci, R., Battistelli, M., Falcieri, E., Agostini, D., Gioacchini, A.M., Stocchi, V. (2011). Creatine as an antioxidant. *Amino Acids, 40,* 1385-1396.

Shimazu, S., Minami, A., Kusumoto, H., Yoneda, F. (2005). Antidepressant-like effects of selegiline in the forced swim test. *Eur. Neuropsychopharmacol., 15,* 563–571.

Silveri, M. M., Parow, A.M., Villafuerte, R.A., Damico, K.E., Goren, J., Stoll, A.L., Cohen, B.M., Renshaw, P.F. (2003). S-adenosyl-L-methionine: effects on brain bioenergetic status and transverse relaxation time in healthy subjects. *Biol. Psychiatry, 54,* 833-839.

Singh, A. (1982). Chemical and biochemical aspects of superoxide radicals and related species of activated oxygen. *Can. J. Physiol. Pharmacol., 60,* 1330-1345.

Sora, I.; Nishimon, K.; Otsuki, S. (1986). Dexamethasone suppression test and noradrenergic function in affective and schizophrenic disorders. *Biol. Psychiatry, 21,* 621-631.

Stone, E.A., Quartermain, D. (1999). Alpha-1-noradrenergic neurotransmission, corticosterone, and behavioral depression. *Biol. Psychiatry, 46,* 1287-1300.

Subramanian, A., Miller, D. M. (2000). Structural analysis of alpha-enolase. Mapping the functional domains involved in down-regulation of the c-myc protooncogene. *J. Biol. Chem., 275,* 5958-5965.

Suzuki, E., Yagi, G., Nakaki, T., Kanba, S., Asai, M. (2001). Elevated plasma nitrate levels in depressive states. *J. Affect. Disord., 63,* 221-224.

Szewczyk, B., Pałucha-Poniewiera, A., Poleszak, E., Pilc, A., Nowak, G. (2012). Investigational NMDA receptor modulators for depression. *Expert Opin. Investig. Drugs*, *21*, 91-102.

Tachikawa, M., Fukaya, M., Terasaki, T., Ohtsuki, S., Watanabe, M. (2004). Distinct cellular expressions of creatine synthetic enzyme GAMT and creatine kinases uCK-Mi and CK-B suggest a novel neuron-glial relationship for brain energy homeostasis. *Eur. J. Neurosci.*, *20*, 144-160.

Tachikawa, M., Hosoya, K., Ohtsuki, S., Terasaki, T. (2007). A novel relationship between creatine transport at the blood-brain and blood-retinal barriers, creatine biosynthesis, and its use for brain and retinal energy homeostasis. *Subcell. Biochem.*, *46*, 83-98.

Trivedi, M.H. (2006). Major depressive disorder: remission of associated symptoms. *J. Clin. Psychiatry, 6*, 27-32.

Trullas, R., Skolnick, P. (1990). Functional antagonists at the NMDA receptor complex exhibit antidepressant actions. *Eur. J. Pharmacol.*, *185*, 1-10.

Trzeciak, H. I., Pudełko, A., Gabryel, B., Małecki, A., Kozłowski, A., Ciesślik, P. (1995). Effect of antidepressants on ATP content, 3H-valine incorporation and cell morphometry of astrocytes cultured from rat brain. *Dev. Neurosci.*, *17*, 292-299.

Tuglu, C., Kara, S.H., Caliyurt, O., Vardar, E., Abay, E. (2003). Increased serum tumor necrosis factor-alpha levels and treatment response in major depressive disorder. *Psychopharmacology (Berl), 170*, 429-433.

Usala, T., Clavenna, A., Zuddas, A., Bonati, M. (2008). Randomised controlled trials of selective serotonin reuptake inhibitors in treating depression in children and adolescents: a systematic review and meta-analysis. *Eur. Neuropsychopharmacol.*, *18*, 62-73.

Van Pilsum, J.F., Stephens, G.C., Taylor, D. (1972). Distribution of creatine, guanidinoacetate and the enzymes for their biosynthesis in the animal kingdom. Implications for phylogeny. *Biochem. J.*, *126*, 325-345.

Venkatraman, T.N., Krishnan, R.R., Steffens, D.C., Song, A.W., Taylor, W.D. (2009). Biochemical abnormalities of the medial temporal lobe and medial prefrontal cortex in late-life depression. *Psychiatry Res.*, *172*, 49-54.

Volz, H. P., Rzanny, R., Riehemann, S., May, S., Hegewald, H., Preussler, B., Hübner, G., Kaiser, W.A., Sauer, H. (1998). ^{31}P magnetic resonance spectroscopy in the frontal lobe of major depressed patients. *Eur. Arch. Psychiatry Clin. Neurosci, 248*, 289-95.

Walker, J. B. (1979). Creatine: biosynthesis, regulation, and function. *Adv Enzymol. Relat. Areas Mol. Biol.*, *50*, 177-242.

Walliman, T., Eppenberger, H. M. (1985). Localization and function of M-line-bound creatine kinase. M-band model and creatine phosphate shuttle. *Cell. Muscle Motil.*, *6*, 239-285.

Williams, M.H., Kreider, R.B., Branch, J.D. (1999). Creatine: The power supplement. Human Kinetics Publishers, 1999.

Wong, M.L., Licinio, J. (2001). Research and treatment approaches to depression. *Nat. Rev. Neurosci*, *2*, 343-351.

World Health Organization (2009). The World Health Report - Mental Health: New Understanding, New Hope.

Wyss, M., Kaddurah-Daouk, R. (2000). Creatine and creatinine metabolism. *Physiol. Rev.*, *80*, 1107-1213.

Yamada, J., Sugimoto, Y., Yamada, S. (2004). Involvement of dopamine receptors in the anti-immobility effects of dopamine re-uptake inhibitors in the forced swimming test. *Eur. J. Pharmacol.*, *504*, 207-211.

Yang, Y., Yang, D., Tang, G., Zhou, C., Cheng, K., Zhou, J., Wu, B., Peng, Y., Liu, C., Zhan, Y., Chen, J., Chen, G., Xie, P. (2013). Proteomics reveals energy and glutathione metabolic dysregulation in the prefrontal cortex of a rat model of depression. *Neuroscience*, *247C*, 191-200.

Young, J. F., Larsen, L.B., Malmendal, A., Nielsen, N.C., Straadt, I.K., Oksbjerg, N., Bertram, H.C. (2009). Creatine-induced activation of antioxidative defence in myotube cultures revealed by explorative NMR-based metabonomics and proteomics. *J. Int. Soc. Sports Nutr. 7*, 9.

Zafir, A., Banu, N. (2007). Antioxidant potential of fluoxetine in comparison to Curcuma longa in restraint-stressed rats. *Eur. J. Pharmacol.*, *572*, 23-31.

Zafir, A., Ara, A., Banu, N. (2009). Invivo antioxidant status: a putative target of antidepressant action. *Prog. Neuropsychopharmacol. Biol. Psychiatry*, *33*, 220-228.

Zarate, C.A., Singh, J.B., Carlson, P.J., Brutsche, N.E., Ameli, R., Luckenbaugh, D.A., Charney, D.S., Manji, H.K. (2006). A randomized trial of an N-methyl-D-aspartate antagonist in treatment-resistant major depression. *Arch. Gen. Psychiatry, 63*, 856-864.

Zomkowski, A.D.E., Rosa, A.O., Lin, J., Santos, A.R., Calixto, J.B., Rodrigues, A.L.S. (2004). Evidence for serotonin receptor subtypes involvement in agmatine antidepressant like-effect in the mouse forced swimming test. *Brain Res*, *1023*, 253-263.

Zomkowski, A.D., Engel, D., Gabilan, N.H., Rodrigues, A.L.S. (2010). Involvement of NMDA receptors and L-arginine-nitric oxide-cyclic guanosine monophosphate pathway in the antidepressant-like effects of escitalopram in the forced swimming test. *Eur. Neuropsychopharmacol.*, *20*, 793-801.

Zomkowski, A.D., Engel, D., Cunha, M.P., Gabilan, N.H., Rodrigues, A.L.S. (2012). The role of the NMDA receptors and L-arginine-nitric oxide-cyclic guanosine monophosphate pathway in the antidepressant-like effect of duloxetine in the forced swimming test. *Pharmacol. Biochem. Behav.*, *103*, 408-417.

In: Creatine
Editors: F. D'Cruz and V. Ribeiro

ISBN: 978-1-62948-305-4
© 2013 Nova Science Publishers, Inc.

Chapter II

Therapeutic Uses of Creatine: New Possibilities

Martina Antošová[1] and Jana Plevková[2]
[1]Department of Physiology,
[2]Department of Patological Physiology
Comenius University Jessenius Faculty of Medicine,
Martin, Slovakia

Abstract

Creatine is an organic acid that contribute to the energy supply to the skeletal muscle. The basic substrates for the creatine biosynthesis in the human body are semi-essential amino acids L-arginine, glycine and methionine. The creatine is synthesized in the liver and kidneys and partially can be substituted from food such as fish or meat. It is an essential source of muscle energy, which is used mainly by athletes to improve short-term performance. The creatine is one of the best-selling nutritional supplements. Effects of the creatine on increasing muscle strength was also tested in pathological conditions and diseases such as injuries or surgeries associated with prolonged immobilization, in patients with muscle weakness associated diseases, in patients with increased fatigue, in some neuromuscular diseases, in patients with muscular dystrophy, amyotrophic lateral sclerosis, chronic obstructive pulmonary disease or in patients with chronic heart failure. The creatine can be applied in various forms - creatine monohydrate, creatine phosphate, creatine maleate, creatine citrate etc. Based on available data we can conclude that the effect depends on the pharmacokinetic parameters of the forms of the creatine used for the supplementation. After per oral supplementation the creatine is rapidly and reversibly phosphorylated in the muscles. This reaction is catalyzed by creatine kinase, which produces creatine phosphate as a basic source of ATP resynthesis. ATP is subsequently used as a basic energy source. Use of the creatine is not introduced yet to standard practice, however some of the experimental and clinical studies indicate that supplementation of the creatine improves function of skeletal muscle in some diseases (such as COPD), and consequently the quality of life of affected patients. The creatine supplementation can replace anabolic steroids or nutritional supplements in conditions associated with malnutrition, such as cancers or COPD. Very interesting areas represent neuroprotective effects of creatine, effects on cognitive and mental functions, influence on the metabolism of carbohydrates and fats. Currently, the use of the creatine is very popular especially among young people and athletes in order to increase muscle strength, muscle growth and weight loss. The creatine may also, as mentioned above, improve the performance especially in relation to short-term sports activity. The results of the creatine supplementation in the pathological conditions remain controversial. The studies reported positive results but the effects of the creatine often depend on many factors like comorbidities and individual metabolic differences. The review summarizes the current therapeutic possibilities of the creatine as well as the results of recent research activity in the creatine field.

Introduction

Regular use of the creatine is one of the factors responsible for improved performance in active athletes. Their muscles must be able to provide maximal effort. To do that, muscle cells need to renew energy resources in adenosine triphosphate (ATP) – lactic acid cycle. However, in case of sufficient concentration of the creatinephosphate they can cover the immediate energy demands by their own storage of it. It means that muscle cells have a hierarchy of energy utilization in which the ATP goes first, then cell utilizes creatinephosphate, and finally the pathway turns to the lactic acid system. Substitution of the creatine may lead to formation of the store of creatine and better renewal of ATP. Extreme physical exercise usually leads to the glycogen depletion therefore supplementation of creatine is frequently combined with the supplementation of carbohydrates. This combination has positive effect on muscles and it leads to faster accumulation of glycogen and muscles performance becomes more effective. Same time it increases the persistence in the strength muscle training and muscle ability to up-take exogenous creatine.

The effect of the creatine and its impact on muscle cells and gain of the muscle mass has been recognized on the tissue cultures in 1976 [1]. This finding considerably influenced commercial synthesis of specific protein systems. British Olympic team used this concept in the training for Olympics in Barcelona in 1992, to increase the performance of UK sportsmen. Based on the data and better physical performance after the creatine treatment, it was broadly used by professional and also recreational athletes to enhance their physical performance and ability [2]. Moreover, the concept of the creatine supplementation became attractive in relation to certain specific diseases characterized by abnormalities in energy pathways, or muscle diseases.

Although the creatine treatment in certain diseases is at the phase of clinical trials, it is nearly impossible to ignore recent research articles that document positive effects of the creatine on muscle performance, muscle mass gain, persistence and improvement of the quality of life, mainly in patients with chronic diseases.

The effects of the creatine on human cells became more and more interesting for research teams in different brands of research therefore there are still some new information up-dating its current effects and possible clinical application of this organic acid. This chapter cannot describe all of the possible targets for the creatine supplementation; however, it could be an inspiration and motivation for further research in this rapidly growing and progressive area of research.

Biosynthesis, Metabolism and Transport of the Creatine

Creatine ($C_4H_9N_3O_2$ - N-(aminoiminomethyl)-N-methyl glycine) is an organic acid, which is synthesized in small amounts in the human body, in the liver and the kidneys. There are recognized three substrates for the creatine synthesis – all of them are amino acids – L-arginine, methionine and glycine. Synthesis of the creatine consists of two steps. The first step is synthesis of L-ornithine and guanidinacetate from the glycine and L-arginine in the kidneys. This pathway is catalyzed by the enzyme arginine-glycine-amidinotransferase

(AGAT). AGAT is the specific enzyme and its activity is the limiting factor in the creatine synthesis. Then guanidinacetate as the end-product of AGAT is the direct precursor for the creatine synthesis [3].

The second step of the creatine synthesis is a methylation, which takes places in the liver. Guanidinacetate-N-methyltransferase (GAMT) releases the methyl-group from the methionine-guanidinacetate via S-adenosylmethionine. GAMT is the enzyme able to bind substrates in certain order, therefore the S-adenosylmethionine is the first substrate and it is followed by the guanidinacetate as the second substrate. Synthesis of the creatine is finalized after transfer of the methyl- group (Figure1) [3].

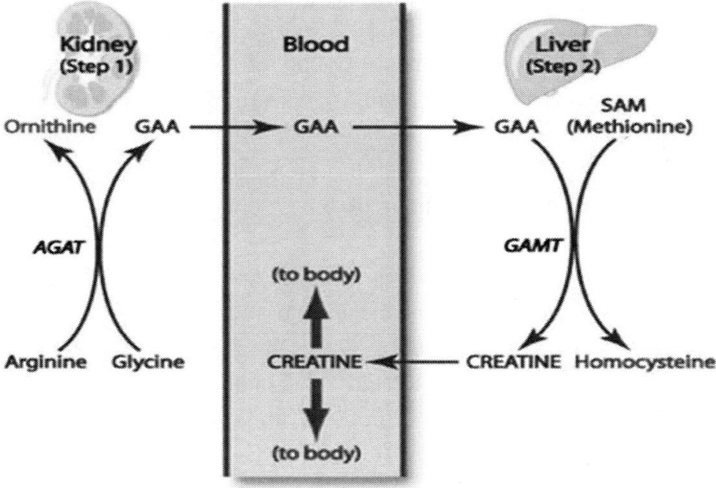

Figure 1. Creatine biosynthesis. (AGAT = Arginine:glycine amidinotransferase; GAMT = Guanidinoacetate methyltransferase GAA= Guanidinoacetic acid) Copyright http://www.creatinemonohydrate.net/creatine_b-vitamins.html.

Once the synthesis is finalized, the creatine is distributed all over the body. Nearly 95% of the creatine is transported to the muscles, the rest of it - 5% is distributed to the brain, retina, kidneys, liver and testicles [3].

Synthesis of the creatine is influenced by the availability of the substrates, synthesizing enzymes, and regulatory factors such are thyroid hormones, testosterone, growth hormone etc. [4]. It could be also influenced by the type of diet, regarding the exogenous creatine intake mainly in the meat and fishes.

The creatine could be substituted directly as semisynthetic or synthetic preparation in a form of nutrition supplements. The creatine demands are in physiological conditions provided by the way of intestinal absorption (from food) or by its synthesis *de novo*. It is interesting that organs with the highest activity of the AGAT and/or GAMT contain paradoxically the lowest concentration of the creatine.

Transportation to the tissues is finalized by creatine up-take at the cell membrane by sodium-potassium channel against the concentration gradient. Similar mechanism is known for dopamine, guanidine, gamma amino-butyric acid (GABA) or taurine [5]. This mechanism operates in the kidneys, myocardium, skeletal muscles, testicles and large intestine, but not in the liver, pancreas and small intestine. The recent data indicate that creatine cannot be transported across the hematoencephalic barrier and the brain has its own independent creatine synthesis [6,7].

The transformation of the creatine to creatine-phosphate runs in the tissues and is catalyzed by the enzyme creatinekinase (CK). Based on the activity of the involved enzymes we can estimate that human body synthesizes approximately 120 grams of the creatine daily.

Main Effects of the Creatine

Creatine is the main substrate with high importance in the metabolism of the energy supplying systems, particularly in the muscles and neurons. It contributes to the regulation of many physiological processes.

Energy Resources

The creatine participates in the production of ATP via its integration to the creatine phosphate energy system. Creatine becomes natural component of the renewal of the energy resources in the cell by the transformation of ADP (adenosin diphosphate) to ATP. Physiological process of the energy release is based on the turn of ATP to ADP by disconnection of the macroergic bond between two phosphates. The creatine then becomes a carrier of the phosphate therefore it becomes also potential donor of the phosphate. This chemical group can be added to the ADP from the creatine phosphate to recharge the energy back to the of ATP, and the creatine becomes free again to carry another phosphate groups or to be metabolized to creatinine, which is in turn eliminated from the cells, transported by blood to the kidneys, where it is a subject for glomerular filtration [8].

Creatine together with creatine phosphate participates in the transport of ATP across the internal mitochondrial membrane to the cytoplasm. It is important to note, that creatine participates in the regulation of the glycolysis. In case of the exhaustion of the organism, tissues containing the creatinin can adapt to this situation by the up-regulation of the activity of certain oxidative enzymes, which in turn may help to the energy sustainability in the cells. These up-regulated enzymes are creatinekinase, citrate synthase, sukcinatedehydrogenase etc. [9].

Creatine phosphate energy system is actively involved to the production of energy resources and their renewal. This knowledge is frequently utilized by thousands of athletes all around the world. Physical exercise represents specific catabolic process, which initiates a lot of metabolic, tissue and hormonal reactions. Physical activity leads to the consumption and further depletion of the energy, and moreover, it may lead to micro traumatism of the muscle cells. Santos et al. (2004) documented that the creatine significantly prevents the leak of the lactate dehydrogenase from traumatized muscles. Supplementation of the creatine improves the energy balance of the cells, and also improves the recovery after exercise.

Regulation of the energy cycle has effect on many physiological processes with or without the age as a determining factor. There is a pool of the data concerning energy metabolism in the ageing. Many of the results obtained in these studies document the positive effect of the creatine supplementation on ageing.

In general, cells of the growing or young organisms have better capabilities to produce and renew energy, to eliminate toxic metabolic endproducts, and they have better antioxidative capabilities. Moreover cells of the young organism have better protective mechanisms against internal and external noxas. These abilities are reduced with ageing. This may lead to accumulation of potentially dangerous substances inside of the cell, activating pathways leading to the cell death. All of the described processes, but mainly processes related to the renewal of the cell functions and structures require energy, and its availability depends on properly functioning mitochondrial systems. Function of the mitochondria declines with the age and this decline can be further modulated by the physical activity. In simple terms older organism operates with less effective energy production and renewal systems. Creatine is important factor in this process. It significantly enhances and fastens the process of energy (ATP) renewal and supports the metabolism of the muscles. Supplementation of the creatine leads to the increase of the strength and performance of the elderly population [11]. The main effects are increased performance and faster muscle regeneration, what may indirectly slow the ageing.

Neuromodulation

The level of the creatine influences the function of the central and peripheral neural system in healthy individuals and also in subjects with neuronal diseases. This is the reason that creatine was assigned as neuromodulatory molecule. It is more or less clear that its neuromodulatory action is related to the capability to interfere with the energy metabolism and creatinekinase/creatin-phosphate system, which is critically important for the cells with the high metabolic demands, such are neurons for example. Damage to the system of the energy production or renewal may lead to the neurodegenerative diseases and other neurological problems related to ageing. The creatine also modulates an appetite and plays a role in the regulation of body weight [6].

Protein Synthesis

Some data from the research indicates that creatine as the end product of the muscle contraction can be responsible for the rise of the amount of contractile elements in the muscle cells. Hespel and colleagues (2001) suggested that supplementation of the creatine increased the diameter of the both types of the muscle fibers. Creatine alone has impact on the synthesis and accumulation of the specific and nonspecific muscle proteins in skeletal muscle and the myocardium. It can also influence the proteosynthesis indirectly, for example by the hyperhydratation of cells. In such circumstances creatine acts as an osmotic agent and the hyperhydratation itself is the anabolic signal initiating the proteosynthesis [13]. At the other side, hypoosmolality can be a signal for the formation of the protein pool, or for the decrease of the protein degradation [14].

Stabilization of the Cell Membrane

The creatine participates in the process of the stabilization of the cell membrane. Creatine phosphate binds to the phospholipids of the membrane thus decreasing the fluidity of it. Same time it prevents potential leak of the cytoplasm including the enzymes, which normally operates inside of the cells for example creatinekinase [4].

Modulation of the Activity of Biological Signals (Hormones)

Growth hormone is important signal not only for the growth of the long bones, but it participates in the regulation of other important processes such are regulation of the fat metabolism, wound healing, modulation of the locomotoric apparatus and its functioning etc. Creatine enhances the release of the growth factor and basically this interaction is believed to be a key factor leading to the growth effect frequently observed after the creatine substitution. This indirect anabolic effect is used not only in sportsmen, but also in elderly population, patients with muscular dystrophy and cachexia from different reasons [15].

Other Effects

The literature sources document also antioxidative and antiapoptotic capabilities of the creatine.

Creatine Substitution

The creatine is used mainly in a form of nutrition supplements in otherwise healthy individuals. Its therapeutic use is underestimated instead there is a pool of relevant evidence that supplementation of the creatine has significant effect on improvement of certain diseases characterized by reduced energy production and renewal. In the last couple of years the creatine became one of the most popular nutrition supplements. It is used mainly in active and recreational sportsmen to enhance the muscle performance and support the muscle gain. Benefits of the supplementation of creatine in healthy individuals was proven and the evidence is clear for improvement of the muscle activity, strength, persistence and performance [16], and a muscle gain, which was measured by the diameter and volume of the muscle fibers [17]. This effect is age dependent, and the effect of the creatine on muscles decreases with the age. This reduced effect of supplemented dose is probably a consequence of changed up-take and absorption pattern, or changed and/or less effective transport mechanisms [18].

The most frequently used preparation for the creatine supplementation is the creatine monohydrate, but the pool of the creatine containing substances for commercial use is growing every year. Application of the new forms of the creatine is linked to the identification and then description of the physical and chemical properties of these substances, their better biological availability, reduced side effects and higher safety profile. Introduction of the new form of the creatine preparations is limited and is under control by the USA, EU and other important markets authorities. At the moment, the creatine monohydrate is the only one preparation which is approved for supplementation strategies in USA, EU, Canada and South Korea [19].

The creatine is usually supplemented in two phases. The first phase, also called loading phase lasts typically 5 to 7 days, and it is characterized by the high doses of the creatine (20 – 30 g of the creatine) divided into 4 doses per a day. The main objective of the loading phase is to obtain maximal saturation of the muscles with the creatine and to maintain stable concentration of the creatine during the entire day. After the loading phase is finalized, it is optimal to continue in the maintenance phase, which is characterized by lower doses of the creatine 3-5g per a day [2]. The dose depends on the body weight of treated individuals, but it should not exceed 10g even in subjects with overweight. The literature data published in this field share conflicting therapeutic strategies and also the doses differ significantly. The regimes and doses are different for specific muscle and nerve disease. The decision about the creatine dose and regime must consider whether it is supplementation in a healthy individual (sportsmen) or the subjects with diseases (muscle diseases, heart failure etc.). The age is also one of the factors that must be taken into consideration.

The best experiences with the creatine supplementation could be found in professional sportsmen. They use the creatine in the small doses to boost the performance, muscle strength, persistence, and muscle gain. The most frequent way of the creatine administration is per oral treatment with pills or the solutions in the water or juices. The creatine supplementation is usually accompanied by the administration of substances with high glycemic index – for example dextrose, mainly in professional sportsmen. The effect of the creatine after per oral administration is influenced by the factors such are absorption in the gastrointestinal tract, splanchnic perfusion and also by the chemical properties of the preparation itself. To avoid these factors, some authors recommend parenteral administration in the forms of the intramuscular injection [19].

Elimination of the creatine from the plasma has two main routes – renal and extrarenal. Main factor governing the renal elimination of the creatinin is the glomerular filtration. Extrarenal elimination from the plasma is in fact the up-take of the creatine to the muscle cells, which is modified by hormonal influences and some other factors, such are insulin and catecholamines. Muscles up-take is also influenced by the physical exercise [4].

The lack of uniform guidelines frequently leads to a discussion about the optimal doses of the creatine, and also the regimes the creatine is supplemented to the body. Also the necessity of the loading phase is the subject to the discussion, so does the maximal duration of the treatment, which should not exceed 6 weeks. After these 6 weeks cycle a break is recommended for 4 weeks.

Interestingly, there is a group of the subjects who do not respond to the exogenous creatine supplementation - non-responders. Subjects with naturally high concentration of the creatine respond less to the treatment than the subjects with the lover concentrations [20]. The example for the low creatine concentration subjects are vegetarians, which do react robustly to the supplementation of even low doses of the exogenous creatine. The effectiveness of the supplementation is influenced by the alcohol intake, long time use of the simple sugars and irregular physical activity.

Forms of the Creatine

Creatine Monohydrate

Creatine monohydrate is considered to be one of the most common forms of the creatine. Its composition represents 88% of the creatine and 12% of water.

It is worldwide the most frequently used for the commercial objectives thus is the most frequently used for the supplementation strategies. Application of the creatine monohydrate increases the muscle phosphocreatine about 10-40% [19]. Acute and also chronic supplementation increases physical performance, mainly when it is linked with the physical activity on the regular basis [21].

The side effects could be the gain of the body weight, because of the muscle gain, what is in sportsmen one of the expected effects. Instead of creatine monohydrate is very popular; it is not that effective as other forms of the creatine. Its potency rises with the supplementation of the carbohydrates. It is important to note, that approximately 20% of the subjects do not respond to this form of the creatine [19].

Table 1. Descripcion of the Creatine Monohydrate

Designation	Creatine content (%)	Indication	Side effects
Creatine monohydrate	87,9	• amyotrophic lateral sclerosis • Parkinson diseases • others neurodegenerative diseases • brain function, cognitive skills • depression • cancer, oncology diseases • diabetes mellitus • hearth ischemia • lung disease, COPD • muscle recovery after heavy exercise, muscular dystrophy • bodybuilding	• nausea, • stomach upset, • dizziness, • weakness, • loose stools, • diarrhea, weight gain, • muscle cramping, • sporadically proteinuria, • in individual cases depression

Creatine Pyruvate

The main characteristic of this preparation is a combination of two active substances – creatine and pyruvate. Pyruvate is the molecule with specific relationship to the energy cycle and in combination with creatine it can boost the amount of available energy, and same time it supports the reduction of the pool of fat stores.

Supplementation of the creatine in a dose 7 g/day during 7 days did not increase the persistence in cyclists, but the dose 7.5 g/day during 5 days enhanced the aerobic metabolic pathways in water sportsmen, which was documented by the lower level of lactic acid [22, 23]. Similar result was obtained by Jäger and colleagues (2008). Synergistic effect of the creatine and pyruvate can reach better effectiveness than the creatine alone, mainly the effect on the persistence is significantly higher.

Creatine Citrate

Creatine citrate is the combination of the creatine and the citric acid. The effect of this combination was not compared to the effect of the creatine monohydrate so far. Short term application of this preparation increased anaerobic working capacity in healthy, physically active females [25] and moved the onset of the neuromuscular fatigue during the ergometry [26]. The concentration of the creatine in this preparation is less about 40% when comparing to the creatine monohydrate.

Creatine Phosphate

This preparation consists of the creatine and three molecules of the phosphate. It is therefore optimal source for the ATP, mainly for the sportsmen and bodybuilders.

Creatine Ethyl Ester – CEE

This preparation is the monohydrate with the ester group. Classical monohydrate increases the muscle strength and persistence, but its effectiveness depends on its ability of the cells to up take the creatine from the blood. Monohydrate has limited absorption and also limited ability to cross the cell membrane.

Esterification increases the ability to move across the cell membranes. Supplementation regime with the CEE does not need the loading phase and also it is not necessary to add the dextrose or other sugars to the treatment, because they may lead to the accumulation of the fat stores. The literature data do not confirm that CEE would have better effects on the muscle performance or the strength, so the evidence about its higher efficacy is not available [27]. Instead it has positive effects on cognitive functions [28].

Creatine Malate

Preparation is composed of the two or three molecules of the creatine with the malate. This chemical bond is destroyed in the stomach after per oral administration, and it releases creatine and malate. Both components are independently and easy absorbed in the gastrointestinal system. While the malate is an important cofactor and catalyzer for the ATP production in the Krebs cycle, creatine is integrated into the creatinephosphate energy renewal system, thus contributing to the ATP re-charges.

Mg^{2+} – Creatine Chelate

Sophisticated technology allows the creatine to be released from the chemical bond in the chelate before the biotransformation to the creatinine. The result of this reaction is relatively high level of available creatine, which could be directly utilized by the muscle cells. Chelate is transported directly to the muscle cells; therefore released creatine is directly available for the renewal of energy. Magnesium, which is a part of this molecule, also improves energetic and anabolic activity. Based on the mechanism of the direct transport to the muscle cell, supplementation of this preparation does not require the loading phase or additional sugar supplementation doses [29].

Creatine Hydrochloride

This preparation belongs to the new brand of the creatine preparations with excellent absorption from the gastrointestinal system. The absorption of this preparation is about 70% better than absorption of the creatine monohydrate, or other products containing the creatine. This ability can considerably modulate the doses and also reduce or completely cut off the necessity of the loading phase [30].

Creatinol

This preparation is the creatine precursor. Intramuscular and intravenous application of creatinol in the form of creatinol − O- phosphate (COP) significantly increases the muscle strength in the hands of tested subjects [31].

At the moment, there are available also other forms of the creatine - creatine alfa-amino butyrate, alfa-ketoglutarate, taurinate, pyroglutamate, ketoisocaproate etc. These preparations contain approximately 50% of the creatine, however their effectiveness is low incomparable to the creatine monohydrate.

Side Effects of the Creatine Supplementation

Literature data do not report any serious side effects of the creatine supplementation.

The most common reactions, which could be considered as side or adverse effects are dysfunction of the gastrointestinal system and the uropoetic system, muscle cramps and liver disorders.

Mild, but very uncomfortable side effects could be flatulence, which is typical for the creatine preparations with limited solubility. This is the reason why per oral application of the creatine must be accompanied by the fluids intake, or to use modern so called micronize forms of the creatine.

Loading phase could be a source of the mild muscle cramps, mainly in the calf muscles, but it is possible to release them by the administration of the magnesium in a dose 150-600 mg. This phenomenon is explained by the drop of the free magnesium inside of the cell, because the newly formed complex of the creatine and phosphate has ability to bind magnesium reducing its free fraction.

Nonspecific liver and renal impairment are discussed, however not confirmed adverse effects of the creatine supplementation. The data obtained in clinical studies suggest that if these organs are functioning properly before the supplementation, they are probably not at the risk of the damage [32].

Frequently discussed question is whether the long term use of the creatine as a food supplement (several years of use) is safe. This is not satisfactorily explained and confirmed, therefore majority of the literature sources recommend to perform the creatine supplementation in the 6 weeks cycles with the 4 weeks breaks.

These breaks are necessary to reach and maintain basal synthesis of the creatine, which could be suppressed by the supplementation of the external creatine. Moreover, long term supplementation without a break can lead to down regulation of the receptor systems which are responsible for the creatine up take from the blood to the cells, it means, that cells may become unable to up-take the creatine from the blood instead of its supplementation.

Overview of the side effects of the creatine supplementation documents that there are identified only creatine side effects, which are not serious and can be easily managed (muscle cramps). Therefore the creatine could be considered as safe enough and its supplementation does not have dangerous effects on the human body [33].

Therapeutic Application of the Creatine

The lack of the creatine in the human body can be inherited or acquired. Inherited disorders are related to the metabolism of the creatine, for example autosomal recessive disorder of the genes encoding the synthesis of the GAMT or AGAT – enzymes which participate on the creatine synthesis, or the mutation of the creatine transporter SLC6A8. The consequences of the creatine lack are mainly neurological disorders – mental retardation, speech disorders and epileptic symptomatology [34]. As it was described previously, creatine is important for the neuronal homeostasis. Neurons are critically dependent on the ATP availability as an energy source, to maintain resting membrane potential, and to discharge action potentials or transduce the signals. Seriousness of the disorder depends on the type of the enzymopathy or the type of the creatine transport limitation.

Secondary lack of the creatine is a natural consequence of the ageing with the reduction of the muscle mass, inflammatory diseases, oncological diseases etc. Supplementation of the creatine in these diseases can share an antioxidative effect of the creatine and also it can contribute to the faster recovery from diseases and after physical exercise.

Supplementation of the Creatine in the Primary Creatine Syndromes

In general, creatine syndromes are very rare disorders, and it is not completely possible to identify their real prevalence [35]. The pool of the literature reported so far 4 cases of the AGAT deficiency and 40 cases of the GAMT deficiency. The main target in subjects with the primary creatine syndromes is the brain. Mentioned enzyme deficiencies lead to a complete lack of the creatine in the brain, which was confirmed by the data from proton magnetic resonance and spectroscopy of the brain (H-MRS). Interestingly the lack of the creatine does not affect or does not manifest in the skeletal muscles or myocardium. The level of creatine in muscles is normal or slightly reduced. Complete lack of the creatine in the muscles has never been described in the literature yet [34].

Except the mental retardation, and the speech disorders, there were identified other variable clinical presentations in subjects with GAMT deficiency, such are muscle hypotony, progressive pyramidal and extrapyramidal symptoms, self-aggression or autism.

Statistically, more frequent is the lack of the creatine transporter SLC6A8, and more than 20 mutations was identified so far for this protein. Deficiency of this transport system was described for the first time in 2001 in a subject with mental retardation. This disorder was later identified also in other children with the mental retardation and severe problems with the speech development. Prevalence of this disorder is relatively high, and it is supposed that it is responsible for approximately 2% of mental retardation related to X chromosome in male population. Approximately 50% of female (heterozygotes) with mutation of the gene for SLC6A8 have problems with learning and they have been diagnosed with abnormal behavior patterns. Some of the female subjects have clinical manifestation similar to male (mental retardation, epileptic symptomatology). At the other side, some of the heterozygotes are absolutely asymptomatic [36]. Creatine syndromes are still not well characterized and therefore many times not diagnosed properly.

The level of the guanidinoacetate is important in the development of the creatine syndromes. The levels of the guanidinoacetate are normal, or slightly elevated in the GAMT deficiency, but the creatine levels are low. Guanidinoacetate accumulates in the tissues and the body fluids and it is not further metabolized to creatine. In AGAT deficiency the levels of the guanidinoacetate are low, but the creatine in the plasma normal and excretion only slightly reduced. This phenomenon is not satisfactorily explained yet. It is supposed that regulation via exogenous creatine could interfere with the creatine pathways in this syndrome. The concentration of the precursor guanidine acetate is normal in the transporter SLC6A8 deficiency. Elevated ratio between creatine/creatinin is frequently observed in the male suffering from this disorder, as the creatinin is more intensively eliminated by the kidneys, and the re- uptake lacks. In heterozygote females is the ration creatine/creatinin normal, or slightly elevated [34, 35].

In primary creatine deficiencies (GAMT and AGAT disorders) the subjects can be treated by complete supplementation of the creatine. The most frequently used is creatine monohydrate. Its substitution is not effective in the transporter SLC6A8 deficiency.

Recommended Doses of the Creatine Monohydrate for Subjects with Creatine Syndromes

Optimal dose or the patients with the GAMT deficiency is 350 mg of the creatine/ kg of body weight/day. The treatment is usually accompanied by other diet measures, for example supplementation of the L-arginine (15 -25 mg/kg/day) and L-Ornithine (100 mg/kg/day). Peroral substitution of the creatine monohydrate increased the level of the creatine in the brain about 70%. The clinical presentation of the patients was significantly improved – with reduced epileptic symptomatology and improvement of the extrapyramidal symptoms. Positive effects could be seen also additional therapy with L arginine, instead the level of the creatine remain low on supplementation [37, 38].

Optimal dose for the patients with the AGAT deficiency is 400 mg/kg/day. Application of the recommended doses was able to increase the brain creatine level about 60% and it also improved the clinical presentation of the patients. In the AGAT deficiency no additional amino acids supplementation is recommended [39].

Patients with the deficiency of the creatine transporter have optimal dose of the creatine monohydrate of 340 mg/kg/day, often in a combination with the L-arginine and glycine. In general, the literature suggests that single therapy with creatine monohydrate alone is not effective in this case of deficiency. Preparation increases the body weight without any significant effects on brain. At the other hand, L-arginine and glycine as the primary substrates for the creatine biosynthesis with extremely high doses of the creatine monohydrate are relatively effective [34, 40].

Supplementation of the Creatine in Neurological Disorders

The creatine has positive effects in patients with neurological disorders. Its supplementation improves the resistance against the ischemic/reperfusion injury and oxidative stress of the neural tissues. One of its effects is also increased production and renewal of energy via increased activity of the mitochondrial system. The creatine can be substituted in the acute phase of the neurological diseases, and also in diseases with chronic progressive clinical course, for example neurodegenerative diseases. Interestingly, supplementation of the creatine is positive prognostic factor even if the supplementation preceded the onset of the neurological problem [41].

Pyramidal cells responsible for the learning and memory located in the hypothalamus reveal high expression of different isoforms of the enzyme creatinekinase [42]. This finding suggest that supplementation of the creatine can influence learning and memory as well. Majority of the papers published in this area document that supplementation of the creatine improved brain functions, and the most significant effect was detected in subjects on vegetarian diet. Vegetarian subjects on creatine dose 5 g/day during six weeks improved cognitive functions, results of the IQ tests and short term memory [43]. Moreover, supplementation of the creatine in a dose 4-5g/day improved the functioning of the brain under normal physiological conditions, and under stress, which is an indirect evidence for the role of the creatine in psychic performance [6].

The creatine mediated neuroprotection was documented in case of acute brain injury (traumatic, or ischemic/reperfusion injury). The creatine in case of traumatic brain injury significantly improved cognitive functions, speech, and self-assistance in comparison to control subjects [44]. The effect of the creatine in acute brain injury probably depends on the speed of the saturation of injured area. Therefore, it is suggested that per oral application is probably too slow and no effective enough. This is the reason, why majority of researchers recommend direct application to the site of the injury [45, 46].

There is reported interesting impact of the creatine on the neurodegenerative diseases. They are characterized by progressive loss of neurons form the one or more areas of the brain. The main cause of these diseases is not completely known, but factors as genetic predisposition, oxidative damage, energy imbalance and mitochondrial dysfunction are frequently discussed as potentially involved in their pathogenesis. Supplementation of the creatine was tried as the therapeutic trial in neurodegenerative diseases such are Parkinson diseases, Alzheimer dementia, Huntington disease and many more [41].

Parkinson disease is idiopathic neurodegenerative disease characterized by the reduction of the brain dopamine levels. The consequence of the improper function of the dopaminergic neurons in extrapyramidal tract is dyskinesia, bradykinesia, muscle rigidity and tremor. It is supposed that except genetic predisposition, these neurons can be influenced by exogenous factors, thus initiating neurodegeneration. One of suggested mechanisms responsible for the loss of the dopaminergic neurons could be disorder of the mitochondrial electron transport chain, leading to the lack of energy, initiating the degeneration via apoptosis of target neurons [47].

At the moment, there is no such effective treatment that would considerably improve the quality of life, or to prevent progression of the disease. With the progression, the neurological symptoms became more severe and disabling. Except the pharmacological modulation of the dopamine levels, for example levodopa, or dopamine agonists (pramipexol, bromokriptin etc.), selegiline, anticholinergic drugs, amantadine, inhibitors of catechol –o-methyltransferase and rehabilitation techniques, there is reported a trial with creatine on experimental animals with the model of Parkinson disease. Application of the creatine in this model showed neuroprotective properties of the creatine [48]. In clinical trials the creatine supplementation improved mood of the patients and supported the strength training [49, 50].

Alzheimer disease is neurodegenerative disease progressively leading to dementia characterized by diffuse atrophy of the cerebral cortex. Atrophy is a consequence of the lack of the neurons and nerve fiber in the cortex and also by the secondary demyelination of the subcortical areas [51]. Experimental data confirm protective effect of the creatine against beta-amyloid toxicity in children and adults. Some of the research results suggest that there could be direct relationship between metabolism of the creatine and Alzheimer disease, as the creatine has neuroprotective effect in the disease models or neuronal cultures after excitotoxic influence of the glutamate or toxic effect of the beta amyloid [52]. Some of the authors speculate that the creatine supplementation can slow the progression of the disease, however, this speculation need to be further clarified.

Huntington disease belongs to the group of autosomal recessive diseases with the onset at the middle age. Clinical presentation is determined by the progressive damage of the brain by the formation of the lesions. The beginning of the diseases is characterized by the presence of unconscious muscle shivers (chorea), which are progressing into the unconscious shakes and rotatory movement. Patient is not able to perform any of the routine activities, what leads to disability and assistance requirement with reduced quality of life. Except of the motoric disability, the patient may complain of the personality changes, with aggressive behaviors attacks, proceeding to the dementia and complete decline of the patient's personality. Recently, there are some theories suggesting, that disorders of the mitochondrial complexes II and III can be the mechanism leading to the lack of the energy, increasing the level of the lactic acid in the brain a reduction of the creatine phosphate in the muscles, which can explain the symptomatology. These changes were described in the subjects with the Huntington disease located at the level of basal ganglia [53]. Substitution of the creatine is the key mechanism leading to the rise of the creatine phosphate in the muscles and decrease of the lactic acid in the brain – which could be the logic strategy to eliminate neurological and muscle problems. Supplementation of the creatine in animals with artificial brain lesions (induced for example by 3 –nitropropionic acid (3NP)) reduced the extension of the lesions in comparison to not treated animals. The animals treated with creatine also had better markers related to the less extensive oxidative stress [54]. Work of Ferrante et al. (2000) and Sheara at al. (2000) document that supplementation of the creatine in the mice (1-2%) of the food weight) was sufficient to prevent the brain atrophy, reduced atrophy of the striatum, the number of anatomic abnormalities (lesions induced by 3NP) and increased the performance of the experimental animals. So far the results suggest that patients with Huntington disease can profit from the long term supplementation of the creatine, although this effect is small.

Amyotrophic lateral sclerosis (ALS) sometimes named Lou Gehrig's disease is rapidly progressing disease of nervous system, affecting primary motoneurons in the brain, brainstem and the spinal cord. Pathogenesis is complex, and the main mechanisms are progressive loss of neurons, which degenerate as a consequence of the mitochondrial defects with the lack of energy. Affected neurons are not able to discharge action potentials and transduce them to the skeletal muscle. Muscles are therefore not functioning properly and become weak and atrophic. The onset of the disease is characterized by clumsiness, stumbles, and problems with gentle manual skills. Later the disease progresses into the dysphagia, speech problems, spasticity of hyperreflexia. Finally the patient is not able to command his muscles, and the inability to command the diaphragm leads to the respiratory failure. The disease does not have an impact on the cognitive functions, therefore the patient can recognize the prognosis of the disease and he usually get depression. It is possible to use rilusol – the drug releasing the glutamate thereby reducing the damage of the motoneurons in subjects with the ALS. Individual physiotherapy, supportive strategies, speech therapy and symptomatic treatment are necessary components of the complex health care. Administration of the creatine in animals with ALS model increased the survival period about 13 and 26 days: animals did not have increased parameters of oxidative stress (3-nitrotyrosine). This study confirmed the hypothesis that creatine is also a potent antioxidant [56]. The creatine also showed protective effects against progressive loss of motoneurons in substantia nigra [57]. Instead of relatively promising effects in animal models, clinical studies with the creatine in subjects with ALS does not show any benefits of the creatine supplementation, however clinical trials are still running. The creatine was also supplemented in the patients with the insomnia, psychic problems like anxiety and depressions, posttraumatic stress disorder etc. In all of these studies supplementation of the creatine improved clinical course of the problem and reduced the magnitude of the symptoms.

Supplementation of the Creatine in Muscle Diseases

Based on the distribution of the creatine to skeletal muscles (~ 95%), supplementation of the creatine can be important in the therapeutic strategies of muscular diseases. In this case supplementation of the creatine in experimental and clinical settings improved the neuromuscular score, and some of the literature sources recommend continuing in creatine supplementation for a long time period, as it improves the disease clinical course and the outcomes. The problem here is a recommended dose, which differs significantly among children and adult subjects, and the optimal doses as well as treatment guidelines are still not well established.

Kley and colleagues (2011) studied effects of the creatine supplementation in the subjects suffering from genetic muscle diseases, leading to the progressive muscle weakness. There is no pharmacological treatment available, and basically the treatment is symptomatic. Supplementation of the creatine in patients with muscle dystrophy increases the muscle strength; however in case of metabolic myopathies the effect of creatine is minimal.

Supplementation of the Creatine in Cardiovascular Diseases

The effects of the creatine on the myocardium are mainly positive, as it was documented in experimental and clinical settings. Experiments showed that creatinephosphate in the animal model of ischemic − reperfusion injury reduced the risk of ventricular arrhythmias and it also reduces the extension of the necrotic areas [59]. Moreover, the creatine influences biosynthesis of the homocysteine, which is considered to be one of the most important risk factor for onset and progression of atherosclerosis. Some literature sources also pointed that level of homocysteine in the blood could be a better predictive factor or indicator of the cardiovascular damage − even better than cholesterol [60]. As it was described previously, the creatine is synthetized by methylation from guanidinoacetate. Processes and metabolic pathways which reduce the availability of the methyl groups (including the synthesis of the creatine) increase the plasma level of homocysteine, which is dangerous for the endothelial surface. Supplementation of the creatine conserves the methyl groups, and reduces the level of homocysteine having indirect protective effect on endothelium and reduces risk of the atherosclerosis [61].

The creatine can be used also in subjects with chronic cardiac insufficiency. Cardiac insufficiency is a clinical entity characterized by inability of the heart to keep optimal cardiac output thus to provide optimal oxygen and nutrients supply to the peripheral tissues. Decrease of the cardiac output initiates compensatory mechanisms, mainly sympathetic system, which is responsible for tachycardia and increased strength of the myocardial contractions to maintain optimal cardiac output. With increased preload or afterload, the tension inside of the myocardial wall may rise, what is the initial stimulus for the myocardial hypertrophy as chronic compensatory mechanism. Hypertrophy reduces the tension in the myocardial wall, however it increases energy demands. So do the tachycardia, and increased strength of contractions. At one side, these compensatory mechanisms are optimizing the cardiac output, but as we described, they are self-limited. It was also recognized, that compromised heart in subjects with cardiac insufficiency has less concentration of the creatine than healthy heart.

Patients with cardiac insufficiency have significantly decreased quality of life. One of the reasons is that this process influences their everyday activities. Patients are tired and complain about muscle weakness, and fatigue. It is important to note, that muscle weakness and fatigue are very early symptoms which appear as the consequence of activated sympathetic system, with vasoconstriction limiting the blood supply to the gastrointestinal system, kidneys and the muscles. Supplementation of the creatine in the subjects with cardiac insufficiency increased the concentration of the creatine phosphate in the cells of the skeletal muscles. This mechanism influenced positively muscle performance, persistence and strength, thus improving the quality of life of studied subjects [62].

Supplementation of the Creatine in Subjects with Respiratory Diseases

Regarding the supplementation of the creatine in respiratory diseases it is important to note chronic obstructive pulmonary disease (COPD). COPD is worldwide a leading cause in respiratory morbidity and mortality. It is characterized by limitation of the expiratory airflow in the peripheral airways (bronchial obstruction), which is not fully reversible. COPD has a character of progressive diseases with deterioration of the respiratory system and decline of respiratory functions. The main cause leading to COPD is smoking, and considerable air pollution in less developed countries. Inhalation of the smoke and other airborne pollutants initiate an inflammatory response, which leads to the excessive mucus production, airway hyper-responsiveness, and airway remodeling. Except of the typical signs and symptoms as cough and breathlessness, the patients with COPD frequently present with the weight loss, muscle atrophy and muscle weakness. This is a consequence of systemic inflammatory response, which is initiated by the pro inflammatory signaling starting in the respiratory system. Deterioration of the muscle function also decreases the quality of life. The force of the COPD experts to improve the quality of life is also related to the improvement of the muscle function by nutrition supplements leading to a muscle gain. These preparations are anabolic steroids, recombinant human growth hormone, appetite stimulants etc. [63]. The work of Fulda et al., 2005 shows that two week long supplementation of the creatine monohydrate in a loading dose 15g daily increased the muscle strength in upper and lower extremities, and also persistence in subjects with moderate to severe COPD. Maintenance dose of the creatine monohydrate 5 g daily supported the result of the lung physiotherapy and further improved the muscle performance.

Other Possible Application of the Creatine

The lack of serious side effects of the creatine supplementation favored this molecule as an interesting target for many more diseases. Available literature data suggest anti-cancer effects of the creatine [65], positive effects in the studies focused on the development of the bone tissue [66], or diabetes mellitus [67].

Conclusion

This short review summarizes and analyzes available literature, supporting positive effects of the creatine in the process of the energy renewal and muscle performance. Relevant literature sources document that supplementation of the creatine supports the muscle gain, persistence and performance rise. This is the main reason why the creatine is one of the favorite nutrition supplements in active professional and also recreational sportsmen. The effect of the creatine on the energy renewal is relevant for the diseases such are chronic obstructive lung disease, some oncologic diseases, heart insufficiency and finally muscle diseases. Particularly interesting and challenging is the effect of the creatine on the energy renewal in the central nervous system where the supplementation of the creatine in experimental and clinical studies provides positive effect even when applied in a long term manner. The main benefit is a minimum of the adverse effects, with respect to nephrotoxicity or hepatotoxicity. Application of the creatine is also beneficial in congenital creatine syndromes characterized by the creatine deficiency with the neurological symptoms and the mental retardation. Neuroprotection and general positive effects of the creatine supplementation depend on the type of the preparation, applied dose, time course of the supplementation and the route of administration. All of these issues remain to be elucidated in further studies either on experimental or clinical basis.

Our review of available literature is not able to cover all of the areas which are investigating the effects of the creatine supplementation and can possible share positive results. At the other side, we do believe, that it can be an inspiration for further research in this field. Instead of some question remain unresolved so far, the creatine is safe nutrition supplement available in fact without limitations. Supplementation of the creatine does not only help sportsmen, but it can considerably improve the quality of life in subjects with serious diseases.

Acknowledgments

This work was supported with project "The increasing of opportunities for career growth in research and development in the medical sciences", co-financed from EU sources.

References

[1]   Morris, G.E.; Piper, M.; Cole, R. (1976). Quantitative changes in creatine kinase isoenzymes during myogenesis in vitro. *Biochemical Society Transaction, 4(6),* 1063–1065.

[2]   Bird, S. P. (2003) Creatine Supplementation and exercise performance: a brief review. *Journal of Sport Science and Medicine, 2,* 123-132.

[3]   Wyss, M.; Kaddurah-Daouk, R. (2000). Creatine and creatinine metabolism. *Physiological Reviews, 80,* 1107-1213.

[4]   Persky, A. M.; Brazeau, G. A. (2001) Clinical Pharmacology of the Dietary Supplement Creatine Monohydrate. *Pharmacological Reviews, 53,* 161 -176.

[5]   Guerrero-Ontiveros, M. L.; Wallimann, T. (1998). Creatine supplementation in health and disease. Effects of chronic creatine ingestion in vivo: Down regulation of the expression of creatine transporter isoforms in skeletal muscle. *Molecular and Cellular Biochemistry, 184,* 427-437.

[6]   Andres, R. H.; Ducray, A. D.; Schlattner, U.; et al. (2008). Functions and effects of creatine in the central nervous system. *Brain Research Bulletin. 76,* 329-343.

[7]   Béard, E.; Braissant, O. (2010) Synthesis and transport of creatine in the CNS: importance for cerebral functions. *Journal of Neurochemistry, 115 (2),* 297-313.

[8]   Bessman, S. P.; Carpenter, C.L. (1985) The Creatine-Creatine Phosphate Energy Shuttle. *Annual Review of Biochemistry, 54,* 831 – 862.

[9]   O'Gorman, E.; Beutner, G.; Wallimann, T.; et al. (1996). Differential effects of creatine depletion on the regulation of enzyme activities and on creatine-stimulated mitochondirial respiration in skeletal muscle, heart, and brain. *Biochimica Biophysica Acta,* 1276(2), 161-170.

[10]   Santos, R.V.; Bassit, R. A.; Caperuto, E.C.; et al. (2004). The effect of creatine supplementation upon inflammatory and muscle soreness markers after a 30km race. *Life Science, 75 (16),* 1917 – 1924.

[11]   Candow, D. G.; Chilibeck, P.D. (2008). Timing of creatine or protein supplementation and resistance training in the elderly. *Applied Physiology, Nutrition and Metabolism, 33(1),* 184 – 190.

[12]   Hespel, P.; Eijnde, B. O.; Leemputte, M. V.; et al. (2001). Oral creatine supplementation facilitates the rehabilitation of didude atrophy and alters

the expression of muscle myogenic factors in humans. *The Journal of Physiology, 536 (2),* 625-633.

[13] Haussinger, D.; Lang, F.; Gerok, W. (1994). Regulation of cell function by the cellular hydratation state. *The American Journal of Physiology, 267,* E343 –E355.

[14] Berneis, K.; Ninnis, R.; Haussinger, D.; et al. (1999). Effects of hyper- and hypoosmolality on whole body protein and glucose kinetics in humans. *American Journal of Physiology, 276,* E188-E195.

[15] Rahimi, R.; Faraji, H.; Vatani, D. S.; et al. (2010). Creatine supplementation alters the hormonal response to resistance exercise. *Kinesiology, 42 (1),* 28 -35.

[16] Earnest, C.P.; Snell, P.G.; Rodriguez, R.; et al. (1995). The effect of creatine monohydrate ingestion on anaerobic power indices, muscular strenght and body composition. *Acta Physiologica Scandinavica, 153,* 207 – 209.

[17] Volek, J. S.; Duncan, N. D.; Mazzetti, S.A.; et al. (1999). Performance and muscle fiber adaptations to creatine supplementation and heavy resistance training. *Medicine and Science - in Sport and Exercise, 31,* 1147 -1156.

[18] Rawson, E. S.; Clarkson, P.M. (2000). Acute creatine supplementation in older men. *International Journal of Sports Medicine, 21,* 71 – 75.

[19] Jäger, R.; Purpura M.; Shao, A.; et al. (2011). Analysis of the efficacy, safety, and regulatory status of novel forms of creatine. *Amino Acid, 40,* 1369 – 1383.

[20] Syrotuik, D. G.; Bell, G. J. (2004) Acute creatine monohydrate supplementation: a descriptive physiological profile of responders vs. non responders. *Journal of Strength and Conditioning Research, 18(3),* 610 – 617.

[21] Kreider, R. B. (2003). Species – specific responses to creatine supplementation. *American Journal of Physiology. Regulatory, Integrative and Comparative Physiology, 285 (4),* R-725-R726.

[22] Morrison, M. A.; Spriet, L. L.; Dyck, D.J. (2000). Pyruvate ingestion for 7-days does not improve aerobic performance in well-trained individuals. *Journal of Applied Physiology, 89 (2);* 549 – 556.

[23] Van Schuylenbergh, R.; Van Leemputte, M.; Hespel, P. (2003). Effects of oral creatine-pyruvate supplementation in cycling performance. *International Journal of Sports Medicine, 24 (2),* 144 – 150.

[24] Jäger, R.; Metger, J.; Lautmann, K.; et al. (2008). The effects of creatine pyruvate and creatine citrate on performance during high intensity exercise. *Journal of the International Society of Sports Nutrition, 5*, 4.

[25] Eckerson, J. M.; Stout, J. R., Moore, G.A.; et al. (2004). Effect of two and five days of creatine loading on anaerobic working capacity in women. *Journal of Strength and Conditioning Research, 18 (1)*, 168 - 173.

[26] Smith, A. E.; Walter, A.A.; Herda, T. J.; et al. (2007). Effects of creatine loading on electromyographic fatigue threshold during cycle ergometry in college-aged women. *Journal of the International Society of Sports Nutrition, 4*, 20.

[27] Spillane, M..; Schoch, R.; Cooke, M.; et al. (2009) The effects of creatine ethyl ester supplementation combined with heavy resistance training on body composition, muscle performance, and serum and muscle creatine levels. *Journal of International Society of Sports Nutrition, 6*, 6.

[28] Ling, J.; Kritikos, M.; Tiplady, B. (2009). Cognitive effects of creatine ethzl ester supplementation. *Behavioral Pharmacology, 20 (8)*, 673 – 679.

[29] Selsby, J. T.; Di`Silvestro, R. A.; Devor, S.T. (2004). Mg2+-creatine chelate and a low-dose creatine supplementation regimen improve exercise performance. *Journal of Strenght and Conditioning Research, 18 (2)*, 311 - 315.

[30] Miller, D.W.; Vennerstrom, J.L.; Faulkner, M.C. (2009) Creatine oral supplementation using creatine hydrochloride salt. *US Patent 7, 608, 641*.

[31] Nicaise, J. (1975). Creatinol-O-phosphate (COP) and muscular performance: a controlled clinical trial. *Current Therapeutical Research, Clinical and Experimental, 17(6)*, 531–534.

[32] Poortmans, J.R.; Francaux, M. (2000). Adverse effects of creatine supplementation: fact or fiction? *Sports Medicine, 30(3)*, 155 - 170.

[33] Bizzarini, E.; De Angelis, L. (2004). Is the use of oral creatine supplementation safe? *The Journal of Sports Medicine and Phisical Fitness, 44*, 411 - 416.

[34] Ipsirolgu-Stockler, I.; Mercimek-Mahmutoglu, S.; Salomons, G.S. (2012). Creatine Deficiency Syndromes. *In: Inborn Metabolic Diseases Diagnosis and Treatment. Springer XXVIII*, 240 – 247.

[35] Nasrallah, F.; Feki, M.; Kaabachi, N. (2010). Creatine and Creatine Deficiency Syndromes: Biochemical and Clinical Aspects. *Pediatric Neurology, 42,* 163 -171.

[36] Hathaway, S.C.; Friez, M.; Limbo, K.; et al. (2009). X-linked creatine transporter deficiency presenting as a mitochondrial disorder. *Journal of Children Neurology, 25 (8),* 1009 – 1012.

[37] Morris, S.M. Jr. (2007). Arginine metabolism: boundaries of our knowledge. *Journal of Nutrition, 137 (6),* 1602S-1609S.

[38] Schulze, A. (2013). Creatine deficiency syndromes. *Handook of Clinical Neurology, 113,* 1837 – 1843.

[39] Battini, R.; Alessandri, M.G.; Leuzzi, V.; et al. (2006). Arginine: glycine amidinotransferase (AGAT) deficiency in a newborn: early treatment can prevent phenotypic expression of the disease. *Journal of Pediatric, 148 (6),* 828 - 830.

[40] Póo-Argüelles, P.; Arias, A.; Vilaseca, M.A., et al. (2006) X-Linked creatine transporter deficiency in two patients with severe mental retardation and autism. *Journal of Inherited Metabolic Diseases, 29(1),* 220 - 223.

[41] Beal, M. F. Neuroprotective effects of creatine. *Amino Acids, 40(5),* 1305 – 1313.

[42] Kaldis, P.; Hemmer, W.; Zanolla, E.; et al. (1996) 'Hot spots' of creatine kinase localization in brain: cerebellum, hippocampus and choroid plexus. *Developmental Neuroscience, 18(5-6),* 542-54.

[43] Rae, C.; Digney, A.L.; McEwan, S.R.; et al. (2003). Oral creatine monohydrate supplementation improves brain performance: a double-blind, placebo-controlled, cross-over trial. *Proceedings Biological Sciences, 270(1529),* 2147 - 2150.

[44] Sakellaris, G.;, Kotsiou, M.; Tamiolaki, M.; et al. (2006). Prevention of complications related to traumatic brain injury in children and adolescents with creatineadministration: an open label randomized pilot study. *The Journal of Trauma, 61(2),* 322 - 329.

[45] Rebaudo, R.; Melani, R.; Carità, F.; et al. (2000). Increase of cerebral phosphocreatine in normal rats after intracerebroventricular administration of creatine. *Neurochemical Research 2000,* 25(11), 1493-5.

[46] Rabachevsky, A.G.; Sullivan, P.G.; Fugaccia, I.; et al. (2003). Creatine diet supplement for spinal cord injury: influences on functional recovery and tissue sparing in rats. *The Journal of Neurotrauma, 20(7),* 659 - 669.

[47] Schapira, A.H.; Cooper, J.M.; Dexter, D.; et al. (1990). Mitochondrial complex I deficiency in Parkinson's disease. *Journal of Neurochemistry, 54,* 823–827.

[48] Matthews, R.T.; Ferrante, R.J.; Klivenyi, P.; et al. (1999). Creatine and cyclocreatine attenuate MPTP neurotoxicity. *Experimental Neurology, 157,* 142–149.

[49] Bender, A.; Koch, W.; Elstner, M.; et al. (2006) Creatine supplementation in Parkinson disease: a placebo-controlled randomized pilot trial. *Neurology, 67,* 1262–1264.

[50] Hass, C.J.; Collins, M.A.; Juncos, J.L. (2007) Resistance training with creatine monohydrate improves upper-body strength in patients with Parkinson disease: a randomized trial, *Neurorehabilitation and Neural Repair, 21,* 107–115.

[51] Bartko, D.; Čombor, I.; Madarász, Š.; et al. (2008). Demencia Alzheimerovho typu. *Via Practica, 10,* 398 – 402.

[52] Brewer, G.J.; Wallimann, T.W. (2000) Protective effect of the energy precursor creatine against toxicity of glutamate and beta-amyloid in rat hippocampal neurons, *Journal of Neurochemistry,* 74, 1968–1978.

[53] Tabrizi S.J.; Workman, J.; Hart, P.E.; et al. (2000) Mitochondrial dysfunction and free radical damage in the Huntington R6/2 transgenic mouse, *Annals of Neurology, 47,* 80–86.

[54] Matthews, R.T.; Yang, L.; Jenkins, B.G; et al. (1998). Neuroprotective effects of creatine and cyclocreatine in animal models of Huntington's disease. *The Journal of Neuroscience, 18,*156–163.

[55] Ferrante, R. J.; Andreassen, O.A.; Jenkins, B.G.; et al. (2000) Neuroprotective effects of creatine in a transgenic mouse model of Huntington's disease, *The Journal of Neuroscience,* 20, 4389–4397.

[56] Shear, D.A.; Haik, K.L.; Dunbar, G.L; (2000) Creatine reduces 3-nitropropionic-acidinduced cognitive and motor abnormalities in rats, Neuroreport, 11, 1833–1837.

[57] Lawler, J.M.; Barnes, W.S.; Wu, G.; et al. (2002) Direct antioxidant properties of creatine. Biochemical Biophysical Research Communication, 290, 47–52.

[58] Kley, R.A., Vorgerd, M.; Tarnopolsky, M.A. (2011). Creatine for treating muscle disorders. *Cochrane Database Systematic Review 24(1),* CD004760.

[59] Rosenshtraukh, L.V.; Anyukhovsky, E.P.; Beloshapko, G.G; et al. (1988). Some mechanisms of nonspecific antiarrhythmic action of

phosphocreatine in acute myocardial ischemia. *Biochemical medicine and metabolic biology.40(3)*, 225-36.

[60] Wyss, M.; Schulze, A. (2002). Health implications of creatine: can oral creatine supplementation protect against neurological and atherosclerotic disease? *Neuroscience;112(2)*, 243-60.

[61] Stead, L.M.; Au, K.P.; Jacobs, R.L., et al. (2001). Methylation demand and homocysteine metabolism: effects of dietary provision of creatine and guanidinoacetate. *American Journal of Physiology. Endocrinology and Metabolism, 281(5)*, E1095 – E1100.

[62] Gordon, A.; Hultman, E.; Kaijser, L.; et. al. (1995) Creatine supplementation in chronic heart failure increases skeletal muscle creatine phosphate and muscle performance. *Cardiovascular Research, 30(3)*, 413 - 418.

[63] van de Bool, C.; Steiner, M.C.; Schols, A.M. (2012) Nutritional targets to enhance exercise performance in chronic obstructive pulmonary disease. *Current Opinion in Clinical Nutrition and Metabolic Care,* 15(6):553-60.

[64] Fuld, J. P.; Kilduff, L. P.; Neder, J. A.; et al. (2005). Creatine supplementation during pulmonary rehabilitation in chronic obstructive pulmonary disease. *Thorax, 60,* 531 – 537.

[65] Patra, S.; Ghosh, A.; Roy, S. S.; et al. (2011). A short review on creatine-creatine kinase system in relation to cancer and some experimental results on creatine as adjuvant in cancer therapy, *Amino Acids, 42(6)*, 2319-30.

[66] de Souza, R.A.; Xavier, M.; da Silva, F.F.; et al. (2011) Influence of creatine supplementation on bone quality in the ovariectomized rat model: an FT-Raman spectroscopy study. *Lasers in Medical Science, 27(2)*, 487-95.

[67] Ročić, B.; Znaor, A.; Ročić, P.; et al. (2011). Comparison of antihyperglycemic effects of creatine and glibenclamide in type II diabetic patients. *Wien Medizinische Wochenschrift, 161(21-22)*, 519-23.

Reviewed by:

Juraj Mokrý Assoc. prof., MD., PhD., Department of Pharmacology, Jessenius Faculty of Medicine, Comenius University, Martin

Monika Kmeťová-Sivoňová Assoc. prof., Mgr., PhD., Department of Medical Biochemistry, Jessenius Faculty of Medicine, Comenius University, Martin

In: Creatine
Editors: F. D'Cruz and V. Ribeiro

ISBN: 978-1-62948-305-4
© 2013 Nova Science Publishers, Inc.

Chapter III

Muscle Ergogenic Effects of Creatine Supplementation in Resistance Exercise Training

Roberto Carlos Burini[1]* *and Edilson Serpeloni Cyrino*[2]
[1]Centre for Nutrition and Exercise Metabolism, Botucatu Medical School, Sao Paulo State University (UNESP), Botucatu (SP), Brazil
[2]Center of Physical Education and Sports, Londrina State University Londrina (PR), Brazil

Abstract

Creatine in its monohydrated form (CrM) is on the top of the list among athletes seeking ergogenic supplements for increasing muscle mass and/or strength. It is proposed that CrM supplementation induces cell swelling which would activate downstream cell-volume signaling cascades and affect muscle-cell metabolism. The up regulated protein involved in osmosensing would result in increases in maximal muscle strength and power, fat-free mass, total body water and body weight. These would happen independently of training but overall CrM supplementation actions usually potentiate concurrent heavy-resistance training effects on muscle. CrM has also a major role on post-exercise

---

* Corresponding author: Email: burini@fmb.unesp.br.

induced damage by slowing down the degeneration phase and accelerating the regeneration phase after high-intensive exercises. As such CrM may provide protective effects via increased phosphocreatine synthesis recovering the ATP levels and maintaining the Ca^{2+}-ATPase pump. Such effects would keep the Ca^{2+} inside the sarcoplasmatic reticulum avoiding Ca^{2+} dependent proteases activation, protein breakdown and higher cell membrane permeability (intracellular components leakage). Traditional CrM supplementation protocols include phases of loading and maintenance. Co-ingestion of CrM with carbohydrate and protein may enhance Cr uptake through insulin action. Although there may be various side effects of Cr supplementation the only clinically significant side effect scientifically proven has been body weight gain. Nevertheless, it is recommended to avoid high dosages of Cr during periods of increased thermal stress as those under high ambient temperature/humidity. Thus Cr supplementation can be considered a safe, ethical, legal, inexpensive and effective nutritional intervention particularly when consumed in conjunction with a resistance training protocol.

1. Creatine as Sports Supplement

Creatine [N-(aminoiminomethyl)-N-methylglycine] is an ingredient commonly found in food, mainly in fish and meat, and is sold as a dietary supplement in markets around the world[1]. Creatine (Cr) is produced endogenously by the body at an amount of about 1 g/d. Synthesis predominantly occurs in the liver, kidneys and to a lesser extended in the pancreas[2]. The average 70 kg young male has a Cr pool of around 120-140 g which varies between individuals[3, 4] depending on the skeletal muscle fiber type and quantity of muscle mass[4, 5]. The remainder of the Cr available to the body is obtained through the diet at about 1 g/d for an omnivorous diet. As Cr is predominantly present in the diet from meats, vegetarians have lower resting Cr levels[6].

Since the early research in 1992 suggested creatine could be an effective ergogenic aid, Cr has become widely used as a dietary supplement by athletes wishing to maximize performance and training gains[7]. Cr is one of the most popular and widely researched natural supplements and has been consistently reported to increase muscle Cr content and improve high-intensity exercise capacity [8]. Cr is used as an ergogenic aid to improve muscle mass, strength and endurance [9].

Creatine monohydrate (CrM) is the most used supplement form but regardless of the form supplementation with Cr it has regularly shown to increase strength, fat-free mass, and muscle morphology with concurrent heavy resistance training more than resistance training (RT) alone[2].

CrM supplementation has a number of biochemical and physiological effects and enhances muscle performance in humans [10]. Short-term CrM supplementation increases muscle force and/or power [11-18], whereas chronic CrM supplementation in conjunction with weight training increases maximal muscle strength and power, fat-free mass, muscle fiber size, total body water and total body weight [19-21] compared to placebo.

Traditional supplementation protocols have been utilized to rapidly increase muscle Cr levels [22] and then maintain these levels. Therefore, the protocols have been divided into a loading and maintenance phase. The loading phase is incorporated to rapidly increase muscle Cr levels by as much as 20% [22]. Muscle Cr levels show increases in just 2 days. The maintenance phase allows for these increased muscle Cr levels to remain for the duration of supplementation [23].

2. Ergogenic Claims

Studies have demonstrated that Cr supplementation is effective for increasing strength, muscular power, lean muscle mass and hydration status [24-28].

2.1. Cell-Swelling Signals

The CrM supplementation has been shown to increase fat-free mass and muscle power output possibly via cell swelling[29]. Studies have found significant increases in total body water in men after both short-term[30] and long-term[19] CrM supplementation, and that this acute increase in fluid volume is limited to the intracellular compartment only[30]. The increase in fat-free mass and total body weight is partly due to fluid retention in myocytes caused by the osmotic potential of high intracellular CrM abundance [19, 31]. These phenotypes effects might be due to either the energy buffering physiochemical attributes of the CrM, cell volume regulation, or CrM supplementation's direct influence on cellular metabolism through changes in gene expression [29]. The raised hypothesis is that CrM supplementation would induce cell swelling which in turn would activate downstream cell-volume-sensitive signaling cascades, and affect cellular metabolism [32].

Short-term CrM supplementation in healthy young men activates genes in the skeletal muscle that are involved in various aspects of osmosensing. CrM supplementation and possible changes in cellular osmolarity significantly up regulated (1.3 to 5.0-fold) the mRNA content of genes and protein contents of kinases involved in osmosensing and signal transduction, cytoskeleton remodeling, protein and glycogen synthesis regulation, satellite cell proliferation and differentiation, DNA replication and repair, RNA transcription control and cell survival [29].

Cell swelling has been identified as an anabolic proliferative signal [33, 34]. A significant increase in protein kinase Ba (PKBa) mRNA and protein content was observed in the skeletal muscle with CrM supplementation (Figure 1). PKBa is a serine-threonine protein kinase and is involved in up regulating general protein synthesis via an increase in the eukaryotic initiation factor 2-B [35]. This findings together with observation that CrM supplementation (for 9 days) decreases whole body protein breakdown and amino acid oxidation (leucine) in healthy young men [32] indicates a trend toward a net myofibrilar protein accretion [29].

In addition to protein synthesis, PKBa positively regulates glycogen synthesis by deactivating glycogen synthase kinase-B, thus increasing glycogen synthase activity [35]. It is suggested that PKBa associates with GLUT-4 containing vesicles, promoting their translocation to the plasma membrane to stimulate glucose transport [36]. However 5 days of dietary Cr supplementation and controlled normal habitual dietary intake had no effect on muscle glycogen storage in healthy male volunteers. Concomitant exercise training might be an important factor. Thus dietary Cr association increases in muscle glycogen content are a result of an interaction between dietary supplementation and other mediators of muscle glucose transport such as muscle contraction (and AMP kinase activation). The major factor influencing the potential for Cr to enhance muscle glycogen content is contraction in the form of glycogen-depleting exercise before Cr supplementation or training during supplementation. In the absence of prior exercise, the ingestion of Cr does not increase muscle glycogen storage [37].

CrM-mediated cell swelling would activated cell volume-sensitive signaling cascades to adapt to the intracellular and extracellular changes in osmolarity by activating the signal transduction pathway for maintaining proper cell function [32]. The suggested mechanism includes cellular swelling due to Cr entry into muscle causing increased glycogen synthase [38], increased muscle Na^+-K^+-ATPase pump activity enhancing Cr transport.

CrM supplementation up regulates genes mRNA with a vital role in osmosensing specially during the earlier phase of cell swelling, by promoting cell adhesion to the extracellular matrix and inducing intracellular signal transduction [39, 40]. Among the modulated genes are those involved in transmembrane signaling systems of hormonal regulation of adenylate cyclase in response to β-adrenergic stimuli [41, 42] and also genes involved in sensing extracellular calcium concentration that assists in maintainin calcium homeostasis [43].

CrM supplementation activates genes involved in cytoskeleton remodeling [29]. The cytoskeleton rearrangements also have been implicated as a result of cell swelling, as an adaptive homeostatic response[44]. One observation is that an increase in mRNA expression of fibrillar collagen genes that plays role in muscle development and cell adhesion[45], with a phospho-protein that regulates reorganization of the actin cytoskeleton and links actin filaments to membrane, glycoproteins by interacting with integrins, transmembrane receptor complexes, and second messengers with a type III intermediate filament cytoskeleton protein that participates in the signal transduction events that play a role in the regulation of cell development, activation and growth [46].

CrM supplementation in conjunction with strength exercise training increases satellite cell number, myonuclei concentration [47] and type II muscle fiber area [21]. Cell swelling may act as an anabolic stimulus facilitating downstream myogenic regulatory factor (MRF) pathways that, in turn, may stimulate satellite cells (post natal myogenic stem cell) to proliferate and fuse with existing myofibers [48] (Figure 1).

Myoblast proliferation and differentiation involve the activation of several key signaling pathways such as ERK 1/2, p38 MAPK and PI3K/PKBa [49]. P38MAPK, a stress-sensitive kinase, is up regulated in response to cell swelling [50]. CrM supplementation induces robust increases in p38 MAPK and PKBa mRNA and protein content which are mandatory for myoblast differentiation and to enhance the expression of downstream myogenic differentiation markers, such as myogenin, myosin heavy chain, and caveolin-3, that may lead to muscle accretion [51]. CrM supplementation increases myogenic transcription factors (MRF4, MEF2A, MEF2C, MEF2D and myogenin) [52].

Thus CrM supplementation activates the transcription of genes that may play a role in downstream cellular proliferation and differentiation in addition to influencing DNA replication and repair, nuclear chromatin organization and anti-apoptotic pathways [29]. It is also proposed that CrM supplementation induces rapid and coordinate induction of regulatory proteins at the molecular level, resulting in increases in maximal muscle strength and power, fat-free mass, total body water and total body weight, independent of training and/or dietary intervention.

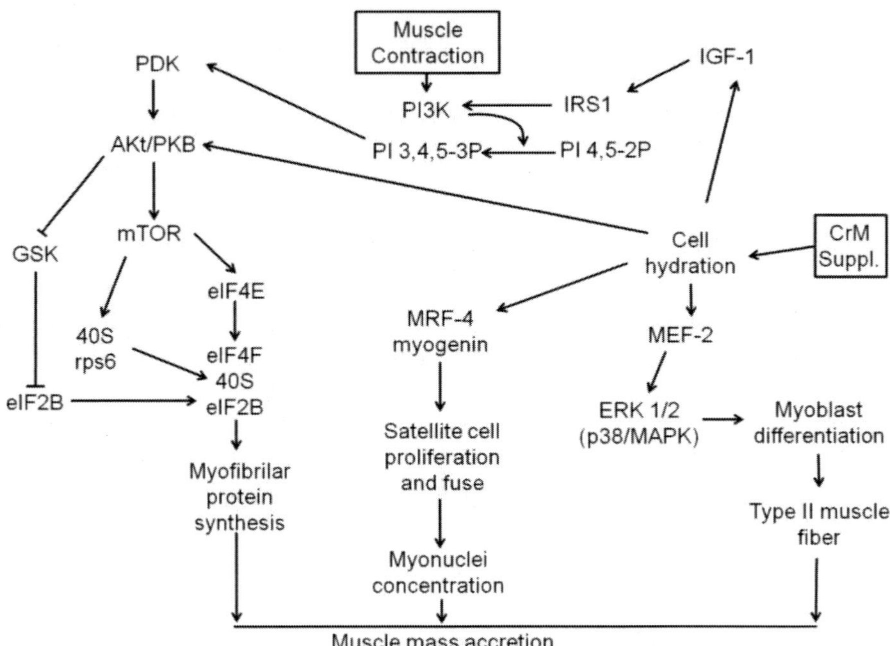

PDK: Phosphoinositide-dependent kinase; PI3K: Phosphoinositol 3 kinase; AKt/PKB: Protein Kinase B; GSK: Glycogen synthase kinase; ERK: Extracellular-signal regulated kinase; p38/MAPK: p38 mitogen-activated protein kinase; mTOR: mammalian target of rapamycin; 40S: ribosomal sub unit 40S; rpS6: ribosomal protein S6; eiF2B: eukaryotic initiation factor 2B, eiF4E: eukaryotic initiation factor 4E; eiF4F: eukaryotic initiation factor 4F; IGF-1: Insulin-like growth factor 1; IRS-1: Insulin Substrate Receptor 1; PI 3,4,5-3P: Phosphatidylinositol 3,4,5 tri phosphate; PI 4,5-2P: Phosphatidylinositol 4,5 bi phosphate; MRF 4: myogenic regulatory factor 4; MEF-2: myocyte enhancer factor-2.

Figure 1. Potential metabolic sites for exogenous creatine actions boostering the exercise effects on hypertrophy.

2.2. Muscular Strength and Power

The combination of CrM supplementation and resistance training (RT) has been shown to synergistically accentuate muscular hypertrophy and muscle cross sectional area [21,48]. Whole-body resistance exercise protocols performed in a flat pyramid loading pattern which includes sets and repetitions with 80-90% of 1-RM is the most effective method to gain maximal strength[9].

The hypertrophic growth occurs in myofibrillar diameter and length by predominant protein synthesis over protein breakdown. Muscle protein synthesis is finely controled by kinase cascade having mTOR as central role; mTOR controls the protein synthesis initiation complex, being positively influenced by muscle contraction, IGF-1 and nutrients [53] (Figure 1).

Cr is an important molecule that participates in buffering intracellular energy stores [54]. The benefits of Cr ingestion have been primarily attributed to the concomitant increase in total muscle phosphocreatine content and a more rapid supply of adenosine triphosphate (11). CrM is a naturraly occurring organic acid that acts as a buffer for cytosolic and mitochondrial pools of adenosine triphosphate (ATP) [55]. Cr is responsible for energy transfer from mitochondria to cytosol; the ATP/ADP ratio maintenance to enhance mitochondria respiration; the attenuation of ADP increase and minimizing the adenine nucleotide loss; and the activation of glucogenolysis and glycolysis through Pi release and by them integrating the carbohydrate and Cr degradation to provide energy at the onset of exercise [56].   Thus one possibility of higher strength gain by Cr supplementation under RT would be due to the acute increase in muscle phosphagen levels, with improved promotion of greater training adaptations (1). When combined with RT, Cr supplementation can increase further muscular strength and isokinetic peak torque, even in the presence of significantly less increase in body volume [57]. No significant differences in body weight, lean body mass and arm muscle area were observed after Cr supplementation and resistance training suggesting that the meaning full increase in muscle strength was promoted without changes in body composition [58]. Evidence accumulated over the past decades suggested that the primary mechanisms by which Cr exerts its anabolic effect in healthy subjects who are weight training is by allowing them to work at a higher proportion of their maximal voluntary contraction force and thus increase the training stimulus [59].

2.3. Protein Synthesis

Exercise is known to activate multiple signal transduction pathways and to modulate transcriptional and translational process [60]. Studies showing modulation of multiple gene clusters in response to resistance exercise stimulus [61-63].

Responses to resistance exercise alone: a) an acute rise in the expression of MAFbx mRNA seen immediately at the end of exercise followed by a subsequent fall at 24h before the return to basal values by 72h; b) a fall in myostatin mRNA at 24h; c) PCNA (a marker of cell proliferation) is activated in a biphasic way, increasing immediately post exercise and after 3 days of recovery; d) an increase in both the sarcoplasmic and nuclear complements of p38 MAPK and ERK 1/2; e) the immediate rise after exercise of MEF-2 protein [60].

It has been suggested that during contractile activity, protein synthesis is depressed and protein breakdown is stimulated. An acute up regulation of MAFbx mRNA (component of the ubiquitin/proteasome pathway) in human muscle immediately after exercise is likely to reflect increased protein breakdown at this time. At 24h post exercise, MAFbx mRNA expression was reversed and depressed, suggesting a reduction in protein degradation. Similar to MAFbx 24h post exercise mRNA for myostatin was depressed (myostatin is a negative regulator of muscle mass and regulates the expression of MAFbx to modulate Ubiquitin-dependent proteolysis [64].

Exercise-induced increases in protein synthesis can last up to 72h [65]. Candidate pathways with resistance exercise, includes protein kinase B (PKB) – mTOR and mitogen-activated protein kinase (MAPK) pathways, key cascades in the regulation of skeletal muscle synthesis and remodeling by resistance exercise [66-70]. ERK 1/2 phosphorylation has been found to increase by more than 50-fold in the sarcoplasm and by about 8-fold in the nucleus (but not statistically significant). ERK 1/2 is another member of the MAPK pathways and like p38 is known to be regulated by exercise [71].

PGC-1α mRNA has been reported to increase after resistance exercise, and the maximal expression level is reached about 3h after exercise [72-74]. This is in agreement with the observation that p38 MAPK is activated and MEF-2 is more abundant in the nucleus immediately after exercise since the transcriptional regulation of PGC-1 α partially depends on the activation of both p38 MAPK and MEF-2. MEF-2 is increased after a bout of resistance exercise [75, 76].

The effects of resistance exercise to PKB cascade is controvertial showing activation [67, 77], no changes [68, 77-79] or decreasing [60]. The decrease in the PKB phosphorylation state fits well with the findings that exercise in the fasted state (usual moment of biopsy) decreases protein synthesis and increases protein breakdown.

Immediately after exercise, IL-6 mRNA more than tripled, confirming that skeletal muscle is major site of IL-6 production [80]. GLUT-4 mRNA abundance decreases by 24h after exercise. Endurance exercise increases GLUT-4 mRNA whereas the opposite effect is observed after resistance exercise [16, 80].

One bout of exercise already stimulates the expression of MHC IIA. Strength training in humans has been shown to result in MHC IIB − to − IIA transitions without affecting MHC I percentage [81]. These changes in the amounts of the different MHC mRNA isoforms precede the corresponding changes at the protein level [82].

After 10 days of Cr supplementation it was observed an increase in fat free mass, total body water and body weight in healthy young men. CrM supplementation increased mRNA expression and protein content of genes involved in protein synthesis regulation [29].

Twelve weeks of Cr supplementation resulted in increased muscle fiber size cross sectional area when compared to placebo for type I, type IIa and type IIb fibers [21]. Additionally, greater increases were found for type I, type IIa and type IIx MHC mRNA expression for Cr supplemented individuals compared to placebo [23].

For the modulation of gene expression by Cr, there has been evidence of increased myosin heavy chain (MHC) and IGF-I and II in human subjects after Cr ingestion [83, 84]. Cr supplementation has been reported to trigger the skeletal muscle expression of insulin like growth factor 1 (IGF-1) to increase fat-free mass [85]. Additionally Cr supplementation during resistance-exercise training increases intramuscular IGF-1 concentration in healthy men and women [6].

The Cr induced increase in collagen 1 (α1) and MHC I-IIa mRNA might improve muscle framework, providing a favorable environment for muscle mass accretion after a few weeks of training. Additional effects of Cr on resistance exercise-induced muscular mass responses are much less pronounced that the effects of exercise alone [60].

2.4. Muscle Damage Repair

Exercise-induced skeletal muscle injury is well understood as the product of unfamiliar or strenuous physical activity and eccentric (lengthening) contractions under high loads [86, 87]. The damage that occurs is primarily due to eccentric muscle actions and affects the structural composition of muscle leading to impairments in performance. Increased muscular strain as a result of the eccentric exercise is believed to cause structural damage within the muscle[23]. The consequences of exercise-induced muscle damage particularly that of eccentric exercise include events that lead to reductions in force, increased soreness, and impaired muscle function [46]. As a result of exercise-induced muscle damage, there is injury to the cell membrane which triggers the inflammatory response leading to the synthesis of prostaglandins and leukotrienes [88].

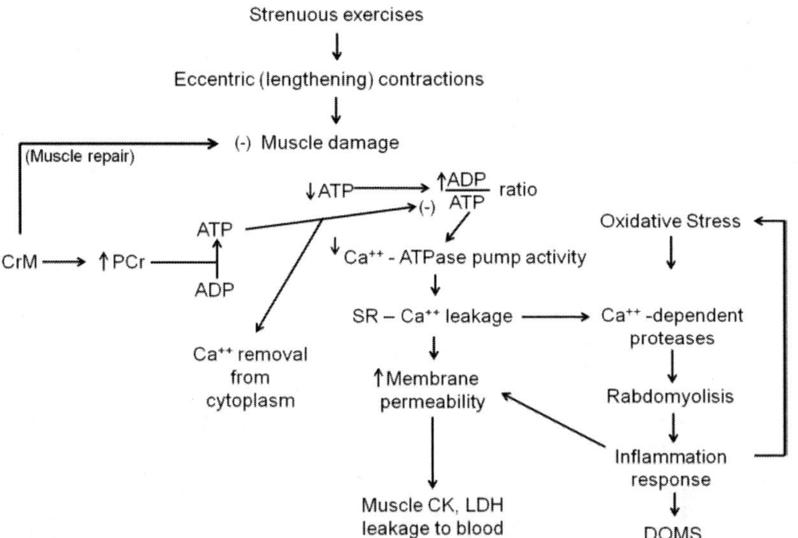

PCr: Phosphocreatine; DOMS: Delayed onset muscle soreness; SR: Sarcoplasmic reticulum; CK: Creatine kinase, LDH: Lactate dehydrogenase.

Figure 2. Potential metabolic sites for exogenous creatine actions on muscle recovery.

The symptoms of delayed onset muscle soreness (DOMS) which include strength loss, pain, muscle tenderness, stiffness and swelling, have been reported to occur within 48 hours of damage and last beyond 5 days. Degradation of contractile proteins appears to contribute to decreases in muscular force 5-28 days post eccentric exercise [89]. Therefore, reductions in force output immediately following a bout of eccentric exercise and up to 5 days may be related to the inflammatory response associated with cellular membrane damage [88]. Decreased muscle force following muscle damage is a result of proteolysis at days 14 and 28 post damage. Early decrements in force (within 5 days of damage) are not attributed to proteolysis [89].

The disruption of normal muscle ultra structure by eccentric exercise alters sarcolemmal and sarcoplasmic reticulum function which results in an increase in intracellular calcium and subsequent activation of degradative pathway [90] associated with muscle degradation. The sarcoplasmic reticulum Ca^{2+} pump derives its ATP preferently from PCr via CK reaction. Local rephosphorylation of ADP by the CK-PCr system maintains a low ADP/ATP ratio within the vicinity of the SR Ca^{2+} pump and ensures optimal Ca^{2+} pump function (i.e., removal calcium from the cytoplasm). However when rates of Ca^{2+} transport are high (as seen in muscle damage), there is a potential for an increase in [ADP], thus creating a microenvironment (i.e., high [ADP]/[ATP] ratio) that is unfavorable for ATPase function, and as a consequence, SR Ca^{2+} pump function may be diminished. Furthermore, a decrease in [PCr] below 5mM, which is characteristic of this increased ATPase activity, reduces local ATP regeneration potential of the CK-PCr system [91] (Figure 2).

Increased CK in blood is a result of eccentric exercise which is unfamiliar to the muscle or muscle group. Additionally the CK elevations in the chronic condition suggest increased sarcolemmal or sarcoplasmic reticulum membrane damage. Increased CK is most likely due to greater sarcolemmal and sarcoplamic reticulum membrane instability as a result of mechanical stress from eccentric exercise [92].

Besides CK, LDH is another myofibrilar enzyme widely accepted as a markers of muscle damage after prolonged exercises [93, 94]. Maximum CK and LDH activity occurs approximately 72 to 96 hours after RT. The delay in maximal elevation of CK and LDH activity is most likely caused by the increasing membrane permeability due to secondary or delayed onset damage as a result of increasing Ca^{2+} leakage into the muscle [95]. The differences in magnitude of CK and LDH present in plasma following eccentric exercise is possible due to larger molecule weight of LDH compared to CK and hence a decreased ability to diffuse from the muscle cell following injury. Less leakage of muscle cell enzymes from the muscle may be indicative of less damage to the muscle, which may be due to improved Ca^{2+} buffering capacity of the muscle (i.e., the rate of Ca^{2+} removal from the muscle cytoplasm) and thus less Ca^{2+}accumulation within the cell and subsequent proteolytic activation [90].

The exercise-induced muscle injury may be prevented or attenuated by adequate training or dietary supplements [96]. One supplement that may reduce the severity of exercise-induced muscle damage or promote recovery is CrM [90]. Athletes have anedoctally reported decreased fatigue, decreased muscle soreness and decreased recovery time while supplementing with Cr with evidence to support increased myofibrillar protein synthesis, muscular hypertrophy and muscular strength with Cr supplementation.

Studies suggested that Cr could be more effective in muscle submitted to the degeneration/regeneration process induced by exercise [47]. Cr would slow down the degeneration phase and accelerate the regeneration phase after high-intensity exercise, which usually occurs one and three days post-exercise, respectively [60].

As a result of exercise-induced muscle damage alterations in sarcolemal and sarcoplasmic reticulum membranes are evident. This damage may result in increased intracellular calcium levels which may be associated with muscle degradation. As such, ingestion of exogenous Cr may provide protective effects via increased phosphocreatine (PCr) synthesis which may aid in stabilizing the sarcolemmal membranes and thereby reducing the extent of damage [92].

It is possible that Cr supplementation will have positive effects on indirect markers of exercise-induced muscle damage [21, 84]. Cr supplementation decreased plasma activities of CK, LDH and aldolase and prevented the rise of AST and ALT plasma activities. Thus Cr supplementation presented a protective effect on muscle injury induced by strenuous contractile activities [96].

Increased muscular strain as a result of the eccentric exercise is believed to cause structural damage within the muscle thereby limiting creatine's effects on cellular membrane stability. Thus 5 days of supplementation (20 g/day) did not reduce indirect markers of muscle damage or reduce recovery time following eccentric exercise. As such, sarcolemmal (and sarcoplasmic) reticulum damage may have been too extensive for a 5 days supplement protocol to have impact [92]. By supplementing with Cr prior to, but also following exercise-induced muscle damage, PCr concentration within the muscle will be increased, and therefore could theoretically improve the intracellular Ca^{2+} handling ability of the muscle by enhancing CK-PCr system and increasing local rephosphorylation of ADP to ATP, thus maintaining a high [ATP]/[ADP] within the vicinity of SR Ca^{2+}-ATPase pump during intense, eccentric exercise [90].

There was a significant improvement in the rate of recovery of knee extensor muscle function after Cr supplementation following injury. Therefore the major finding seemed to be the significantly higher muscle strength after Cr supplementation during recovery from a muscle damaging exercise session [90].

Cr supplementation has been found to enhance intramuscular adaptations to strength training both at the fiber and molecule level [21, 84]. It has been suggested that by continuing Cr supplementation after a resistance exercise bout (initial stimulus), Cr may act as a co-regulator, or direct manipulator of gene transcription of amino acid pools, thus enhancing myofibrillar protein synthesis during the recovery period post injury [97].

The ergogenic effect of Cr following 30 days of supplementation may have a positive impact on exercise induced muscle damage. Maximal isometric force (MIF) following chronic Cr conditions resulting in greater values versus placebo suggests that the Cr supplementation conditions may reduce the extent of muscle damage when supplementing for more than 30 days. The resultant muscle force differences in chronic condition support an ergogenic benefit. Based on enhanced Cr uptake with exercise and positive effects on skeletal muscle composition longer supplementation protocols may be helpful in decreasing exercise-induced damage [23].

The MIF differences between Cr and placebo condition at 30 days post damage may be explained by enhanced protein synthesis versus degradation, most likely within the MHC [84]. Thus in the event that there is increased protein synthesis and therefore reduced proteolysis, then Cr supplementation may have a positive influence on performance when muscle damage occurs [23].

Some contradictory data are available. For example, maximal isometric force of the elbow flexors and serum CK and LDH activity, in response to eccentric exercise were not significantly different between the Cr-supplemented and control groups during the 5 days following exercise [92]. Therefore, it was suggested that Cr supplementation does not reduce indirect markers of muscle damage or enhance recovery from high-force eccentric exercise [90]. Contrarily this isometric (21% higher) and isokinetic (10% higher) knee extension strength were both significantly greater during recovery with consumption of a Cr-CHO supplement compared to a supplement with CHO alone. Such varying responses in the magnitude of strength loss following eccentric exercises are possibly due to the different muscle groups used and/or the protocol utilized to induces muscle damage [90].

Short-term Cr supplementation did not appear to attenuate the effects of exercise induced muscle damage when compared to placebo treatments. However the long-term effects appeared to have had an ergogenic effect on muscle when muscles were subjected to isometric force development. The possibility exists that creatine's ergogenic effects on muscle may require greater than 7 days to passively impact muscle damage [23].

Cr perhaps works via two possible mechanisms. First Cr supplementation prior to eccentric induced damage may be enhancing the Ca^{2+} buffering capacity of the muscle by fueling the SR Ca^{2+}-ATPase pump thereby decreasing intracellular Ca^{2+}concentrations and activation of degradative pathways such as calpain (Figure 2). Second, Cr supplementation post-exercise may enhance one or more of the phases involved in the regenerative response to exercise-induced damage, such as increasing protein synthesis, reducing protein degradation and thus creating an environment that facilitates enhanced satellite cell proliferation and hence formation of new muscle fibers [90].

2.4.1. Antioxidant

The resistance exercise session is designed to be hypoxic in nature, as opposed to high force. Aerobic exercises cause whole-body oxygen consumption, which increases 10-to-20 fold over the resting state and can result in elevated levels of free radicals [98]. Despite the lower oxygen demands compared to those of aerobic exercise, resistance exercise generation of free radicals through other mechanisms is possible: a) xanthine-xanthine oxidase pathway; b) respiratory burst of neutrophils; c) catecholamine auto oxidation; d) local muscle ischemia-hypoxia, and; e) conversion of the weak superoxide to the strong hydroxyl radical by lactic acid [99]. The ROS cause extensive DNA damage including single-stand breaks and the formation of modified bases. One of the most abundant forms of oxidized DNA, 8-hydroxy-2-deoxyguanosine (8-OHdG) represents 5% of the total oxidized bases of the DNA. Acute resistance exercise can lead to acute oxidative stress [100]. An increase in blood malonyldialdehyde (MDA) a marker of lipid peroxidation, was reported in 2 days and immediately after a full-body resistance-exercise protocol [101].

Antioxidant properties of Cr may be attributed to the presence of arginine (Arg) in its molecule. Arg is also substrate for the nitric oxidase synthase family and can increase the production of nitric oxide, a free radical that modulates metabolism, contractility and glucose uptake in skeletal muscle [31, 102].

Cr was first reported by Lawler et al. [103] as capable of directly quenching aqueous radical and reactive species ions. Myotubes have their antioxidant mechanisms stimulated when exposed to CrM [85]. Oxidative stress and oxidative DNA damage were lesser in a Cr supplemented group than in placebo group after a single bout of resistance exercise [9]. Short-term Cr supplementation was able to decrease these oxidative damage of DNA and lipid peroxidation in trained subjects, probably because of increased the activity of antioxidant enzymes and reducing oxidant production. However others had shown that short-term Cr supplementation did not enhance non-enzymatic antioxidant defenses or protect against lipid peroxidation induced by exhaustive exercise [58].

While data suggest that Cr supplementation causes a significant increase in athletic performance by attenuating oxidative stress damage [9], others state that Cr supplementation might induce oxidative stress and decreases total antioxidant status [58].

2.5. Muscle Regeneration

CrM supplementation alone [52, 104] or in combination with progressive strength training induces satellite cell (post-natal myogenic stem cells) proliferation and differentiation. This effect may be mediated, at least in part, by CrM-induced cell swelling [29] (Figure 1).

Cr supplementation amplifies the training-induced increase in satellite cell number and myonuclei concentration, and thus potentially, muscle regeneration [90]. The activation, proliferation and differentiation of satellite cells are mechanisms by which Cr might increase muscle mass after several weeks of supplementation combined with training [60]. Cr has been shown to have a positive effect on satellite cell activity [47, 104, 105] and muscle protein kinetics [32, 48, 52, 106].

The gene expression changes either in the satellite cells or in the myofiber after each exercise session could be one mechanism by which the induction of satellite cell nuclei into the myofiber occurs after several weeks of resistance training, when coupled with Cr supplementation [47]. CrM supplementation in conjunction with strength exercise training increases satellite cell number, myonuclei concentration [47] and type II muscle fiber area [21]. Cr supplementation has been shown to amplify the increase in satellite cell number and myonuclei concentration in human skeletal muscle fibers during 4-16 weeks of resistance training. These Cr-induced adaptations were associated with an enhanced muscle fiber growth in response to strength training [47].

Exercise did acutely increase a marker of satellite cell proliferation (PCNA, a protein involved in DNA replication maximally expressed in S phase) but there was no additional effect of Cr. Also there was no change in MyoD mRNA (key regulator of muscle remodeling associated with cell proliferation) by creatine [60].

Although the most remarkable effects of Cr are seen after several weeks of supplementation, some of them could be initiated at the transcription level as soon as after 5 days of supplementation. Nevertheless these changes are rather small and might not be related to activation of satellite cell and to muscle mass accumulation observed when Cr is combined with resistance training [47].

Cr supplementation may have a favorable effect on muscle mass and strength independently of exercise [107]. Cr *per se* is able to modify the expression of several genes and that exercise training and/or a process of muscle degeneration/regeneration are not an essential pre requisite [60].

Significant gene alterations observed in response to Cr supplementation include increases in the expression of GLUT-4, collagen 1 (α 1), and MHC I mRNA at rest and increased expression of MHC IIa mRNA immediately post exercise [60]. In addition to the modulation of gene expression, Cr supplementation alone has been shown to augment the increase in satellite cells and also has been shown to augment the increase in satellite cells and myonuclei number induced by several weeks of strength training [47].

3. Forms in the Market

CrM is considered a popular dietary supplement that is used by athletes to increase muscle mass and strength and improve sports performance [102, 107, 108].

Creatine monohydrate, first marketed in the early 1990s, is the form most commonly found in dietary supplement/food products and most frequent cited in scientific literature [1]. CrM is not degraded during normal digestion and that nearly 99% of orally ingested CrM is taken up by muscle or excreted in urine [1].

New forms of Cr are marketed with claims of improved physical, chemical and physiological properties in comparison to CrM. Claims include improved stability when combined with other ingredients or in liquids, improved solubility in water, improved bioavailability and even an increase in performance [1].

The commercial CrM product might contain additional dietary constituents that have been shown not to provide any additional response to changes in body mass, body composition, maximal strength and power output beyond solely attributed to CrM [109]. The primary purpose of these additional dietary constituents (i.e. dextrose, α-lipoic acid, ascorbic acid, etc) with CrM supplementation was to enhance muscle total Cr content gains in the short-term compared with the ingestion of CrM alone [110]. Creatine ethyl ester (CEE) is alleged to increase Cr bio-avalilability. When compared to CrM, CEE was not as effective at increasing serum and muscle Cr levels or in improving body composition, muscle mass, strength and power [111]. Similarly co-ingestion of D-pinitol with Cr has been reported to enhance Cr uptake. However adding D-pinitol to CrM does not appear to facilitate further physiological adaptations while resistance training [112].

Co-ingestion of CrM with various nutrients (e.g., carbohydrate, protein) may enhance Cr uptake to a greater degree (1). Low-dose Cr combined with protein supplementation increases lean tissue mass and results in a great relative increase in bench press but not leg press strength [113].

Cr ingested in combination with relatively small quantities of essential amino acids, simple sugars and protein would stimulate insulin release and augment whole-body Cr retention to the same extent as a large bolus of single sugars. It was demonstrated that muscle total Cr content could be augmented by insulin in humans under physiological conditions − that is when Cr consumed in conjunction with a large quantity of simple sugars (~ 95g). The marked increase in plasma insulin concentration that occurs under these conditions is the stimulus for the increase in muscle accumulation probably as a result of insulin stimulating Na^+-dependent muscle Cr transport [7].

4. Safety

When Cr gained media attention many adverse events were attributed to its use, including the deaths of 3 National Collegiate Athletic Association Wrestlers in 1997. However, autopsy results determined that exertional heat stroke, not Cr, was responsible for these deaths. Speculation that Cr may have influenced exertional heat stroke has resulted in an examination of its role in exercise heat intolerance and hydration status [114].

Besides the described occurrence of side effects of Cr supplementation [115], the only clinically significant side effect reported in literature has been weight gain [1]. A round table sponsored by ACSM in 1990s advised athletes to avoid Cr supplementation if they were wishing to control weight or subjected to strenuous exercise and/or hot environments. Others also recommended avoiding high dosages of Cr during periods of increased thermal stress, such as sports activities performed under high ambient temperature/humidity [114].

Cr as an oral supplement is considered safe and ethical, the perception of safety cannot be guaranteed, especially when administered for long period of time to different populations (athletes, sedentary, patient, active, young or elderly [2]. Thus Cr supplementation is a safe practice when administered within the recommended criteria [10, 25, 109, 116-118]. Nevertheless a possible unexpected outcome related to CrM ingestion is the amount of contaminants present that may be generated during the industrial production [119].

In overall, Cr supplementations is considered to be a safe, inexpensive and effective nutritional intervention particularly when consumed in conjunction with a resistance training regime [120].

5. Legality

The legal and regulatory status of different forms of Cr is uncertain. To date, with exception of Japan, CrM is the only form of Cr to be officially approved or accepted in key markets such as the USA, European Union, Canada and South Korea. In Brazil the Brazilian National Sanitary Surveillance Agency (or ANVISA) recently approved the use of CrM in foodstuffs for athletes (reference "Regulations Concerning Foodstuffs for Athletes". Chapter III, Article 10. The National Health Surveillance Agency Collegiate Board of Directors. 2010). Thus Brazil is the only Latin American country to approve the use of Cr and this approval applies specifically to CrM [1].

Key points: Potential muscle sites for the exogenous creatine ergogenesis

1. Cytoskeleton remodeling
a) ↑ hydration status
b) ↑ α1 collagen

c) myofibrillar protein accretion
  • ↑ myosin heavy chain
  • ↑ type II muscle fiber

d) ↑ satellite cell number   • ↑ myonuclei concentration
2. Increasing force and/or power
a) buffering energy stores
b) ↑ glycogen synthesis
3. Reducing the extent of muscle damage
a) ↑ Ca^{++} buffering capacity

b) slowing down the degeneration phase
  • antioxidant capacity
  • ↓ proteolysis

c) accelerating the regeneration phase

Acknowledgments

- No potential conflicts of interest to this chapter were reported
- Edilson Serpeloni Cyrino: Researched data and critical final reading
- Melvin H Williams: Contributed to discussion and reviewed/edited the manuscript
- Rodrigo M Manda: Contributed to the arts of graphics, format organizations and typewritings
- Roberto Carlos Burini: Extracted data and wrote the manuscript.

References

[1]   Jager, R; Purpura, M; Shao, A; Inoue, T; Kreider, RB. Analysis of the efficacy, safety, and regulatory status of novel forms of creatine. *Amino Acids*, 2011, 40, 1369-83.

[2] Cooper, R; Naclerio, F; Allgrove, J; Jimenez, A. Creatine supplementation with specific view to exercise/sports performance: an update. *J Int Soc Sports Nutr*, 2012, 9, 33.

[3] Bemben, MG; Lamont, HS. Creatine supplementation and exercise performance: recent findings. *Sports Med*, 2005, 35, 107-25.

[4] Brosnan, JT; da Silva, RP; Brosnan, ME. The metabolic burden of creatine synthesis. *Amino Acids*, 2011, 40, 1325-31.

[5] Persky, AM; Brazeau, GA. Clinical pharmacology of the dietary supplement creatine monohydrate. *Pharmacol Rev*, 2001, 53, 161-76.

[6] Burke, DG; Candow, DG; Chilibeck, PD; et al. Effect of creatine supplementation and resistance-exercise training on muscle insulin-like growth factor in young adults. *Int J Sport Nutr Exerc Metab*, 2008, 18, 389-98.

[7] Pittas, G; Hazell, MD; Simpson, EJ; Greenhaff, PL. Optimization of insulin-mediated creatine retention during creatine feeding in humans. *J Sports Sci*, 2010, 28, 67-74.

[8] Jagim, AR; Oliver, JM; Sanchez, A; et al. A buffered form of creatine does not promote greater changes in muscle creatine content, body composition, or training adaptations than creatine monohydrate. *J Int Soc Sports Nutr*, 2012, 9, 43.

[9] Rahimi, R. Creatine supplementation decreases oxidative DNA damage and lipid peroxidation induced by a single bout of resistance exercise. *J Strength Cond Res*, 2011, 25, 3448-55.

[10] Terjung, RL; Clarkson, P; Eichner, ER; et al. American College of *Sports Med*icine roundtable. The physiological and health effects of oral creatine supplementation. *Med Sci Sports Exerc*, 2000, 32, 706-17.

[11] Balsom, PD; Soderlund, K; Sjodin, B; Ekblom, B. Skeletal muscle metabolism during short duration high-intensity exercise: influence of creatine supplementation. *Acta Physiol Scand*, 1995, 154, 303-10.

[12] Birch, R; Noble, D; Greenhaff, PL. The influence of dietary creatine supplementation on performance during repeated bouts of maximal isokinetic cycling in man. *Eur J Appl Physiol Occup Physiol*, 1994, 69, 268-76.

[13] Bosco C, Tihanyi J, Pucspk J, et al. Effect of oral creatine supplementation on jumping and running performance. *Int J Sports Med*, 1997, 18, 369-72.

[14] Brose, A; Parise, G; Tarnopolsky, MA. Creatine supplementation enhances isometric strength and body composition improvements

following strength exercise training in older adults. *J Gerontol A Biol Sci Med Sci*, 2003, 58, 11-9.

[15] Dawson, B; Cutler, M; Moody, A; Lawrence, S; Goodman, C; Randall, N. Effects of oral creatine loading on single and repeated maximal short sprints. *Aust J Sci Med Sport*, 1995, 27, 56-61.

[16] Greenhaff, PL; Casey, A; Short, AH; Harris, R; Soderlund, K; Hultman, E. Influence of oral creatine supplementation of muscle torque during repeated bouts of maximal voluntary exercise in man. *Clin Sci (Lond)*, 1993, 84, 565-71.

[17] Mihic, S; MacDonald, JR; McKenzie, S; Tarnopolsky, MA. Acute creatine loading increases fat-free mass, but does not affect blood pressure, plasma creatinine, or CK activity in men and women. *Med Sci Sports Exerc*, 2000, 32, 291-6.

[18] Volek, JS; Kraemer, WJ; Bush, JA; et al. Creatine supplementation enhances muscular performance during high-intensity resistance exercise. *J Am Diet Assoc*, 1997, 97, 765-70.

[19] Kutz, MR; Gunter, MJ. Creatine monohydrate supplementation on body weight and percent body fat. *J Strength Cond Res*, 2003, 17, 817-21.

[20] Vandenberghe, K; Goris, M; Van Hecke, P; Van Leemputte, M; Vangerven, L; Hespel, P. Long-term creatine intake is beneficial to muscle performance during resistance training. *J Appl Physiol*, 1997, 83, 2055-63.

[21] Volek, JS; Duncan, ND; Mazzetti, SA; et al. Performance and muscle fiber adaptations to creatine supplementation and heavy resistance training. *Med Sci Sports Exerc*, 1999, 31, 1147-56.

[22] Harris, RC; Soderlund, K; Hultman, E. Elevation of creatine in resting and exercised muscle of normal subjects by creatine supplementation. *Clin Sci* (Lond), 1992, 83, 367-74.

[23] Rosene, J; Matthews, T; Ryan, C; et al. Short and longer-term effects of creatine supplementation on exercise induced muscle damage. *Journal of Sports Science and Medicine*, 2009, 8, 89-96.

[24] Branch, JD. Effect of creatine supplementation on body composition and performance: a meta-analysis. *Int J Sport Nutr Exerc Metab*, 2003, 13, 198-226.

[25] Buford, TW; Kreider, RB; Stout, JR; et al. International Society of Sports Nutrition position stand: creatine supplementation and exercise. *J Int Soc Sports Nutr*, 2007, 4 (6), 1-8.

[26] Kreider, RB. Dietary supplements and the promotion of muscle growth with resistance exercise. *Sports Med*, 1999, 27, 97-110.

[27] Rawson, ES; Volek, JS. Effects of creatine supplementation and resistance training on muscle strength and weightlifting performance. *J Strength Cond Res*, 2003, 17, 822-31.

[28] Tipton, KD; Ferrando, AA. Improving muscle mass: response of muscle metabolism to exercise, nutrition and anabolic agents. *Essays Biochem*, 2008, 44, 85-98.

[29] Safdar, A; Yardley, NJ; Snow, R; Melov, S; Tarnopolsky, MA. Global and targeted gene expression and protein content in skeletal muscle of young men following short-term creatine monohydrate supplementation. *Physiol Genomics*, 2008, 32, 219-28.

[30] Ziegenfuss, TN; Lowery, LM; Lemon, PW. Acute fluid volume changes in men during three days of creatine supplementation. *J Exerc Physiol*, 1998, 1, 1-14.

[31] Kraemer, WJ; Volek, JS. Creatine supplementation. Its role in human performance. *Clin Sports Med*, 1999, 18, 651-66, ix.

[32] Parise, G; Mihic, S; MacLennan, D; Yarasheski, KE; Tarnopolsky, MA. Effects of acute creatine monohydrate supplementation on leucine kinetics and mixed-muscle protein synthesis. *J Appl Physiol*, 2001, 91, 1041-7.

[33] Pasantes-Morales, H; Lezama, RA; Ramos-Mandujano, G; Tuz, KL. Mechanisms of cell volume regulation in hypo-osmolality. *Am J Med*, 2006, 119, S4-11.

[34] Ritz, P; Salle, A; Simard, G; Dumas, JF; Foussard, F; Malthiery, Y. Effects of changes in water compartments on physiology and metabolism. *Eur J Clin Nutr*, 2003, 57 Suppl 2, S2-5.

[35] Hajduch, E; Litherland, GJ; Hundal, HS. Protein kinase B (PKB/Akt)--a key regulator of glucose transport? *FEBS Lett*, 2001, 492, 199-203.

[36] Foran, PG; Fletcher, LM; Oatey, PB; Mohammed, N; Dolly, JO; Tavare, JM. Protein kinase B stimulates the translocation of GLUT4 but not GLUT1 or transferrin receptors in 3T3-L1 adipocytes by a pathway involving SNAP-23, synaptobrevin-2, and/or cellubrevin. *J Biol Chem*, 1999, 274, 28087-95.

[37] Ferrante, RJ; Andreassen, OA; Jenkins, BG; et al. Neuroprotective effects of creatine in a transgenic mouse model of Huntington's disease. *J Neurosci*, 2000, 20, 4389-97.

[38] Danieli-Betto, D; Germinario, E; Esposito, A; et al. Sphingosine 1-phosphate protects mouse extensor digitorum longus skeletal muscle during fatigue. *Am J Physiol Cell Physiol*, 2005, 288, C1367-73.

[39] Schliess, F; Reissmann, R; Reinehr, R; vom Dahl, S; Haussinger, D. Involvement of integrins and Src in insulin signaling toward autophagic proteolysis in rat liver. *J Biol Chem*, 2004, 279, 21294-301.

[40] vom Dahl, S; Schliess, F; Reissmann, R; et al. Involvement of integrins in osmosensing and signaling toward autophagic proteolysis in rat liver. *J Biol Chem*, 2003, 278, 27088-95.

[41] Bastepe, M; Gunes, Y; Perez-Villamil, B; Hunzelman, J; Weinstein, LS; Juppner, H. Receptor-mediated adenylyl cyclase activation through XLalpha(s), the extra-large variant of the stimulatory G protein alpha-subunit. *Mol Endocrinol*, 2002, 16, 1912-9.

[42] Chen, M; Gavrilova, O; Liu, J; et al. Alternative Gnas gene products have opposite effects on glucose and lipid metabolism. *Proc Natl Acad Sci U S A*, 2005, 102, 7386-91.

[43] Hendy, GN; D'Souza-Li, L; Yang, B; Canaff, L; Cole, DE. Mutations of the calcium-sensing receptor (CASR) in familial hypocalciuric hypercalcemia, neonatal severe hyperparathyroidism, and autosomal dominant hypocalcemia. *Hum Mutat*, 2000, 16, 281-96.

[44] Pedersen, SF; Hoffmann, EK; Mills, JW. The cytoskeleton and cell volume regulation. *Comp Biochem Physiol A Mol Integr Physiol*, 2001, 130, 385-99.

[45] Imamura, Y; Scott, IC; Greenspan, DS. The pro-alpha3(V) collagen chain. Complete primary structure, expression domains in adult and developing tissues, and comparison to the structures and expression domains of the other types V and XI procollagen chains. *J Biol Chem*, 2000, 275, 8749-59.

[46] Thompson, HS; Scordilis, SP; Clarkson, PM; Lohrer, WA. A single bout of eccentric exercise increases HSP27 and HSC/HSP70 in human skeletal muscle. *Acta Physiol Scand*, 2001, 171, 187-93.

[47] Olsen, S; Aagaard, P; Kadi, F; et al. Creatine supplementation augments the increase in satellite cell and myonuclei number in human skeletal muscle induced by strength training. *J Physiol*, 2006, 573, 525-34.

[48] Hespel, P; Op't, Eijnde, B; Van Leemputte, M; et al. Oral creatine supplementation facilitates the rehabilitation of disuse atrophy and alters the expression of muscle myogenic factors in humans. *J Physiol*, 2001, 536, 625-33.

[49] Wu, Z; Woodring, PJ; Bhakta, KS; et al. p38 and extracellular signal-regulated kinases regulate the myogenic program at multiple steps. *Mol Cell Biol*, 2000, 20, 3951-64.

[50]	Niisato, N; Post, M; Van, Driessche, W; Marunaka, Y. Cell swelling activates stress-activated protein kinases, p38 MAP kinase and JNK, in renal epithelial A6 cells. *Biochem Biophys Res Commun*, 1999, 266, 547-50.

[51]	Donati, C; Meacci, E; Nuti, F; Becciolini, L; Farnararo, M; Bruni, P. Sphingosine 1-phosphate regulates myogenic differentiation: a major role for S1P2 receptor. *Faseb J*, 2005, 19, 449-51.

[52]	Ingwall, JS. Creatine and the control of muscle-specific protein synthesis in cardiac and skeletal muscle. *Circ Res*, 1976, 38, I115-23.

[53]	Manda, RM; Maesta, N; Burini, RC. Metabolic basis of muscle growth. *Rev Bras Fisiol Exerc*, 2010, 9 (1), 52-7.

[54]	Adhihetty, PJ; Beal, MF. Creatine and Its Potential Therapeutic Value for Targeting Cellular Energy Impairment in Neurodegenerative Diseases. *Neuromolecular Med*, 2008, 10 (4), 275-90.

[55]	Hammett, ST; Wall, MB; Edwards, TC; Smith, AT. Dietary supplementation of creatine monohydrate reduces the human fMRI BOLD signal. *Neurosci Lett*, 2010, 479, 201-5.

[56]	Gualano, B; Artioli, GG; Poortmans, JR; Lancha, Junior, AH. Exploring the therapeutic role of creatine supplementation. *Amino Acids*, 2010, 38, 31-44.

[57]	Souza-Junior, TP; Willardson, JM; Bloomer, R; et al. Strength and hypertrophy responses to constant and decreasing rest intervals in trained men using creatine supplementation. *J Int Soc Sports Nutr*, 2011, 8, 17.

[58]	Percario, S; Domingues, SP; Teixeira, LF; et al. Effects of creatine supplementation on oxidative stress profile of athletes. *J Int Soc Sports Nutr*, 2012, 9, 56.

[59]	Kreider, RB; Almada, AL; Antonio, J; et al. ISSN exercise & sport nutrition review: research and recommendations. *Sport Nutr Rev J*, 2004, 1, 1-44.

[60]	Deldicque, L; Atherton, P; Patel, R, et al. Effects of resistance exercise with and without creatine supplementation on gene expression and cell signaling in human skeletal muscle. *J Appl Physiol*, 2008, 104, 371-8.

[61]	Bickel, CS; Slade, J; Mahoney, E; Haddad, F; Dudley, GA; Adams, GR. Time course of molecular responses of human skeletal muscle to acute bouts of resistance exercise. *J Appl Physiol*, 2005, 98, 482-8.

[62]	Psilander, N; Damsgaard, R; Pilegaard, H. Resistance exercise alters MRF and IGF-I mRNA content in human skeletal muscle. *J Appl Physiol*, 2003, 95, 1038-44.

[63] Yang, Y; Creer, A; Jemiolo, B; Trappe, S. Time course of myogenic and metabolic gene expression in response to acute exercise in human skeletal muscle. *J Appl Physiol*, 2005, 98, 1745-52.

[64] McFarlane, C; Plummer, E; Thomas, M; et al. Myostatin induces cachexia by activating the ubiquitin proteolytic system through an NF-kappaB-independent, FoxO1-dependent mechanism. *J Cell Physiol*, 2006, 209, 501-14.

[65] Miller, BF; Olesen, JL; Hansen, M; et al. Coordinated collagen and muscle protein synthesis in human patella tendon and quadriceps muscle after exercise. *J Physiol*, 2005, 567, 1021-33.

[66] Cuthbertson, DJ; Babraj, J; Smith, K; et al. Anabolic signaling and protein synthesis in human skeletal muscle after dynamic shortening or lengthening exercise. *Am J Physiol Endocrinol Metab*, 2006, 290, E731-8.

[67] Dreyer, HC; Fujita, S; Cadenas, JG; Chinkes, DL; Volpi, E; Rasmussen, BB. Resistance exercise increases AMPK activity and reduces 4E-BP1 phosphorylation and protein synthesis in human skeletal muscle. *J Physiol*, 2006, 576, 613-24.

[68] Eliasson, J; Elfegoun, T; Nilsson, J; Kohnke, R; Ekblom, B; Blomstrand, E. Maximal lengthening contractions increase p70 S6 kinase phosphorylation in human skeletal muscle in the absence of nutritional supply. *Am J Physiol Endocrinol Metab*, 2006, 291, E1197-205.

[69] Karlsson, HK; Nilsson, PA; Nilsson, J; Chibalin, AV; Zierath, JR; Blomstrand, E. Branched-chain *amino acids* increase p70S6k phosphorylation in human skeletal muscle after resistance exercise. *Am J Physiol Endocrinol Metab*, 2004, 287, E1-7.

[70] Williamson, D; Gallagher, P; Harber, M; Hollon, C; Trappe, S. Mitogen-activated protein kinase (MAPK) pathway activation: effects of age and acute exercise on human skeletal muscle. *J Physiol*, 2003, 547, 977-87.

[71] Widegren, U; Ryder, JW; Zierath, JR. Mitogen-activated protein kinase signal transduction in skeletal muscle: effects of exercise and muscle contraction. *Acta Physiol Scand*, 2001, 172, 227-38.

[72] Coffey, VG; Shield, A; Canny, BJ; Carey, KA; Cameron-Smith, D; Hawley, JA. Interaction of contractile activity and training history on mRNA abundance in skeletal muscle from trained athletes. *Am J Physiol Endocrinol Metab*, 2006, 290, E849-55.

[73] Pilegaard, H; Saltin, B; Neufer, PD. Exercise induces transient transcriptional activation of the PGC-1alpha gene in human skeletal muscle. *J Physiol*, 2003, 546, 851-8.

[74] Vissing, K; Andersen, JL; Schjerling, P. Are exercise-induced genes induced by exercise? *Faseb J*, 2005, 19, 94-6.

[75] Akimoto, T; Pohnert, SC; Li, P; et al. Exercise stimulates Pgc-1alpha transcription in skeletal muscle through activation of the p38 MAPK pathway. *J Biol Chem*, 2005, 280, 19587-93.

[76] Wright, DC; Han, DH; Garcia-Roves, PM; Geiger, PC; Jones, TE; Holloszy, JO. Exercise-induced mitochondrial biogenesis begins before the increase in muscle PGC-1alpha expression. *J Biol Chem*, 2007, 282, 194-9.

[77] Creer, A; Gallagher, P; Slivka, D; Jemiolo, B; Fink, W; Trappe, S. Influence of muscle glycogen availability on ERK1/2 and Akt signaling after resistance exercise in human skeletal muscle. *J Appl Physiol*, 2005, 99, 950-6.

[78] Coffey, VG; Zhong, Z; Shield, A; et al. Early signaling responses to divergent exercise stimuli in skeletal muscle from well-trained humans. *Faseb J*, 2006, 20, 190-2.

[79] Deshmukh, A; Coffey, VG; Zhong, Z; Chibalin, AV; Hawley, JA; Zierath, JR. Exercise-induced phosphorylation of the novel Akt substrates AS160 and filamin A in human skeletal muscle. *Diabetes*, 2006, 55, 1776-82.

[80] Pedersen, BK; Ostrowski, K; Rohde, T; Bruunsgaard, H. The cytokine response to strenuous exercise. Can *J Physiol Pharmacol*, 1998, 76, 505-11.

[81] Adams, GR; Hather, BM; Baldwin, KM; Dudley, GA. Skeletal muscle myosin heavy chain composition and resistance training. *J Appl Physiol*, 1993, 74, 911-5.

[82] Jaschinski, F; Schuler, M; Peuker, H; Pette, D. Changes in myosin heavy chain mRNA and protein isoforms of rat muscle during forced contractile activity. *Am J Physiol*, 1998, 274, C365-70.

[83] Deldicque, L; Louis, M; Theisen, D; et al. Increased IGF mRNA in human skeletal muscle after creatine supplementation. *Med Sci Sports Exerc*, 2005, 37, 731-6.

[84] Willoughby, DS; Rosene, J. Effects of oral creatine and resistance training on myosin heavy chain expression. *Med Sci Sports Exerc*, 2001, 33, 1674-81.

[85] Young, JF; Larsen, LB; Malmendal, A; et al. Creatine-induced activation of antioxidative defence in myotube cultures revealed by explorative NMR-based metabonomics and proteomics. *J Int Soc Sports Nutr*, 2010, 7, 9.

[86] Kendall, B; Eston, R. Exercise-induced muscle damage and the potential protective role of estrogen. *Sports Med*, 2002, 32, 103-23.

[87] Belcastro, AN; Shewchuk, LD; Raj, DA. Exercise-induced muscle injury: a calpain hypothesis. *Mol Cell Biochem*, 1998, 179, 135-45.

[88] Connolly, DA; Sayers, SP; McHugh, MP. Treatment and prevention of delayed onset muscle soreness. *J Strength Cond Res*, 2003, 17, 197-208.

[89] Ingalls, CP; Warren, GL; Armstrong, RB. Dissociation of force production from MHC and actin contents in muscles injured by eccentric contractions. *J Muscle Res Cell Motil*, 1998, 19, 215-24.

[90] Cooke, MB; Rybalka, E; Williams, AD; Cribb, PJ; Hayes, A. Creatine supplementation enhances muscle force recovery after eccentrically-induced muscle damage in healthy individuals. *J Int Soc Sports Nutr*, 2009, 6, 13.

[91] Duke, AM; Steele, DS. Mechanisms of reduced SR Ca(2+) release induced by inorganic phosphate in rat skeletal muscle fibers. *Am J Physiol Cell Physiol*, 2001, 281, C418-29.

[92] Rawson, ES; Gunn, B; Clarkson, PM. The effects of creatine supplementation on exercise-induced muscle damage. *J Strength Cond Res*, 2001, 15, 178-84.

[93] Schwane, JA; Buckley, RT; Dipaolo, DP; Atkinson, MA; Shepherd, JR. Plasma creatine kinase responses of 18- to 30-yr-old African-American men to eccentric exercise. *Med Sci Sports Exerc*, 2000, 32, 370-8.

[94] Lavender, AP; Nosaka, K. Changes in fluctuation of isometric force following eccentric and concentric exercise of the elbow flexors. *Eur J Appl Physiol*, 2006, 96, 235-40.

[95] Gissel, H; Clausen, T. Excitation-induced Ca(2+) influx in rat soleus and EDL muscle: mechanisms and effects on cellular integrity. Am *J Physiol Regul Integr Comp Physiol*, 2000, 279, R917-24.

[96] Bassit, RA; Pinheiro, CH; Vitzel, KF; Sproesser, AJ; Silveira, LR; Curi, R. Effect of short-term creatine supplementation on markers of skeletal muscle damage after strenuous contractile activity. *Eur J Appl Physiol*, 2010, 108, 945-55.

[97] Allen, DG; Whitehead, NP; Yeung, EW. Mechanisms of stretch-induced muscle damage in normal and dystrophic muscle: role of ionic changes. *J Physiol*, 2005, 567, 723-35.

[98] Konig, D; Wagner, KH; Elmadfa, I; Berg, A. Exercise and oxidative stress: significance of antioxidants with reference to inflammatory, muscular, and systemic stress. *Exerc Immunol Rev*, 2001, 7, 108-33.

[99] Ji, LL. Free radicals and antioxidants in exercise and sports. In: Exercise and Sport Science. W.E. Garrett and D.T. Kirkendall, eds. Philadelphia, PA: Lippincott Williams and Wilkins, 2000. 299–317.

[100] Bloomer, RJ; Goldfarb, AH. Anaerobic exercise and oxidative stress: a review. *Can J Appl Physiol*, 2004, 29, 245-63.

[101] Ramel, A; Wagner, KH; Elmadfa, I. Plasma antioxidants and lipid oxidation after submaximal resistance exercise in men. *Eur J Nutr*, 2004, 43, 2-6.

[102] Reid, MB. Invited Review: redox modulation of skeletal muscle contraction: what we know and what we don't. *J Appl Physiol*, 2001, 90, 724-31.

[103] Lawler, JM; Barnes, WS; Wu, G; Song, W; Demaree, S. Direct antioxidant properties of creatine. *Biochem Biophys Res Commun*, 2002, 290, 47-52.

[104] Vierck, JL; Icenoggle, DL; Bucci, L; Dodson, MV. The effects of ergogenic compounds on myogenic satellite cells. *Med Sci Sports Exerc*, 2003, 35, 769-76.

[105] Dangott, B; Schultz, E; Mozdziak, PE. Dietary creatine monohydrate supplementation increases satellite cell mitotic activity during compensatory hypertrophy. *Int J Sports Med*, 2000, 21, 13-6.

[106] Willoughby, DS; Rosene, JM. Effects of oral creatine and resistance training on myogenic regulatory factor expression. *Med Sci Sports Exerc*, 2003, 35, 923-9.

[107] Johnston, AP; Burke, DG; MacNeil, LG; Candow, DG. Effect of creatine supplementation during cast-induced immobilization on the preservation of muscle mass, strength, and endurance. *J Strength Cond Res*, 2009, 23, 116-20.

[108] Eckerson, JM; Stout, JR; Moore, GA; Stone, NJ; Nishimura, K; Tamura, K. Effect of two and five days of creatine loading on anaerobic working capacity in women. *J Strength Cond Res*, 2004, 18, 168-73.

[109] Kreider, RB; Ferreira, M; Wilson, M; et al. Effects of creatine supplementation on body composition, strength, and sprint performance. *Med Sci Sports Exerc*, 1998, 30 (1), 73-82.

[110] Burke, DG; Chilibeck, PD; Parise, G; Tarnopolsky, MA; Candow, DG. Effect of alpha-lipoic acid combined with creatine monohydrate on human skeletal muscle creatine and phosphagen concentration. *Int J Sport Nutr Exerc Metab*, 2003, 13, 294-302.

[111] Spillane, M; Schoch, R; Cooke, M; et al. The effects of creatine ethyl ester supplementation combined with heavy resistance training on body

composition, muscle performance, and serum and muscle creatine levels. *J Int Soc Sports Nutr*, 2009, 6, 6.

[112] Kerksick, CM; Wilborn, CD; Campbell, WI; et al. The effects of creatine monohydrate supplementation with and without D-pinitol on resistance training adaptations. *J Strength Cond Res*, 2009, 23, 2673-82.

[113] Candow, DG; Little, JP; Chilibeck, PD; et al. Low-dose creatine combined with protein during resistance training in older men. *Med Sci Sports Exerc*, 2008, 40, 1645-52.

[114] Lopez, RM; Casa, DJ; McDermott, BP; Ganio, MS; Armstrong, LE; Maresh, CM. Does creatine supplementation hinder exercise heat tolerance or hydration status? A systematic review with meta-analyses. *J Athl Train*, 2009, 44, 215-23.

[115] Karolkiewicz, J; Szczesniak, L; Deskur-Smielecka, E; Nowak, A; Stemplewski, R; Szeklicki, R. Oxidative stress and antioxidant defense system in healthy, elderly men: relationship to physical activity. *Aging Male*, 2003, 6, 100-5.

[116] Kreider, RB; Wilborn, CD; Taylor, L; et al. ISSN exercise & sport nutrition review: research & recommendations. *J Int Soc Sports Nutr*, 2010, 7, 7.

[117] Schilling, BK. Creatine supplementation and health variables: a retrospective study. *Med Sci Sports Exerc*, 2001, 33 (2), 183-88.

[118] Poortmans, JR; Kumps, A; Duez, P; Fofonka, A; Carpentier, A; Francaux, M. Effect of oral creatine supplementation on urinary methylamine, formaldehyde, and formate. *Med Sci Sports Exerc*, 2005, 37 (10), 1717-20.

[119] Bizzarini, E; De Angelis, L. Is the use of oral creatine supplementation safe? *J Sports Med Phys Fitness*, 2004, 44, 411-6.

[120] Dalbo, VJ; Roberts, MD; Lockwood, CM; Tucker, PS; Kreider, RB; Kerksick, CM. The effects of age on skeletal muscle and the phosphocreatine energy system: can creatine supplementation help older adults. *Dyn Med*, 2009, 8, 6.

In: Creatine
Editors: F. D'Cruz and V. Ribeiro

ISBN: 978-1-62948-305-4
© 2013 Nova Science Publishers, Inc.

Chapter IV

Experimental Evidence that Creatine Supplementation during Pregnancy is Protective for the Neonate

David W. Walker PhD, DSc,[1,2,*] *Hayley Dickinson PhD,*[1]
Stacey Ellery,[1] *Domenic LaRosa,*[1] *Zoe Ireland PhD,*[3]
Syed Baharom,[1,4] *Richard Harding PhD, D.Sc,*[4]
and Rod Snow PhD[5]

[1]The Ritchie Centre, Monash Institute for Medical
Research, Clayton, Melbourne, Australia
[2]Department of Obstetrics and Gynaecology, Monash Medical
Centre, Monash University, Clayton, Victoria, Australia
[3]Queensland Cerebral Palsy and Rehabilitation Research Centre,
Royal Brisbane and Women's Hospital, Herston, Queensland, Australia
[4]Department of Anatomy and Developmental Biology,
Monash University, Clayton, Victoria, Australia
[5]Centre for Physical Activity and Nutrition Research,
School of Exercise and Nutrition Sciences, Deakin
University, Burwood, Victoria, Australia

[*] Corresponding author: David W. Walker, The Ritchie Centre, Monash Institute of Medical
Research, 27-31 Wright St., Clayton, Melbourne, Australia, 3168. Tel.: 613-9594 5372;
Fax: 613 – 9594 6249.

Abstract

The ergogenic and neuroprotective properties of creatine are now well recognised, and creatine supplementation is used widely by healthy adults and as a therapy for patients with neurological and musculo-skeletal conditions where depletion of ATP might be prevented by increasing the intracellular creatine pool. Babies can be born under conditions where poor placental perfusion, intrauterine infection, or depressed respiratory drive result in oxygen deprivation, increasing the risk of damage arising in the brain and other key organs of the neonate due to the inability of cells to rapidly replenish ATP via the creatine kinase pathway. This chapter reviews recent experimental work in pregnant and newborn animals that shows the benefits of creatine supplementation for survival after birth when oxygen and nutrient supply may limit cellular respiration, providing the basis for further consideration of the use of creatine in human pregnancy.

Keywords: Fetus, neonate, hypoxia, organ damage, brain injury, neuroprotection

Abbreviations

CBF	cerebral blood flow
CNS	central nervous system
CrT	creatine transporter
NTP	nucleoside triphosphate
PCr	phosphocreatine

Introduction

There is growing evidence for the use of creatine as a therapy for protecting tissues against injury, particularly the brain (Wallimann et al., 2011). The potential for creatine to be used in pregnancy to benefit the fetus and newborn is reviewed here. Increasing the cellular pool of creatine/ phosphocreatine (PCr) through dietary supplementation or subcutaneous injection has been shown to ameliorate the progress of CNS deterioration in several animal models of neurodegenerative disorders (Adcock et al., 2002a,

Berger et al., 2004b). Creatine has also been shown to decrease the severity of damage in models of acquired brain injury (Wyss and Schulze, 2002). Clinical trials of dietary creatine supplementation in several of these conditions have yielded positive results (Ohtsuki et al., 2002, Schulze, 2003). The key findings of these studies are summarised in Table 1.

One of the primary mechanisms of injury arising from severe hypoxia at birth, particularly for the brain, includes mitochondrial dysfunction, leading to impaired energy metabolism and oxidative stress (Calvert and Zhang, 2005, Wyss and Schulze, 2002). As outlined below, it is now clear that perinatal hypoxia arising: (i) during pregnancy and resulting in fetal growth retardation; (ii) at birth due to problems with placental or umbilical cord perfusion; or (iii) immediately after birth due to problems with the initiation of breathing by the neonate - can result in compromise of several organ systems including the brain, kidney, heart, and musculo-skeletal systems. The ability of exogenously administered creatine to counter the effects of chronic or acute hypoxia on the fetus at birth are discussed below. When the creatine supplementation is given to the mother, a further key question arises of what effects this might have on maternal tissues during pregnancy and the postpartum.

Animal Models of Pregnancy in Relation to Creatine

Consideration must be given to the suitability of various species for the study of creatine supplementation in relation to pregnancy, and the assessment of fetal and neonatal health. Creatine has been shown to transfer from the maternal to fetal compartment in the human, rat, and the spiny mouse (Ireland et al., 2008, Miller, 1974, Miller et al., 1977), but not in sheep where administration of creatine to the ewe does not result in measurable changes of creatine in fetal blood (unpublished observations). As the sheep is a herbivore and also does not absorb creatine from the gut (Xue et al 1989), this raises the possibility that the placental transfer of creatine is restricted to carnivore and omnivore species where creatine is also obtained from the diet.

Table 1. Neuroprotective effects of creatine in neurodegenerative diseases and acquired CNS injury

Disorder	Model/Species	Creatine Treatment	Neuroprotective Outcomes	Reference
Huntington's disease	Transgenic (R6/2) mouse	1-3% oral creatine initiated after onset of symptoms	Delayed development of neuronal shrinkage, gross brain atrophy and formation of Huntington aggregates. Improved survival, rotarod performance and reduced weight loss.	(Dedeoglu et al., 2003, Ferrante et al., 2000)
	3-NP toxicity in rats	1% oral creatine 2 weeks before 3-NP injections	Reduced the decrease of brain energy metabolites, markers of oxidative stress, striatal lesion, and ventricular enlargement. Improved performance on balance-beam task, and working and reference memory.	(Matthews et al., 1998, Shear et al., 2000)
	Humans	Oral creatine ranging from 8 g/d for 16 weeks to 10 g/d for 2 years	Reduced markers of oxidative injury to DNA in serum and unresolved brain glutamate. Prevented weight loss, and although motor and neurological symptoms did not improve over 2 years, some patients showed stabilization of these symptoms.	(Bender et al., 2005, Hersch et al., 2006, Tabrizi et al., 2005)
Parkinson's disease	MPTP-induced dopamine depletion in rats	1% oral creatine 2 weeks before MPTP-injections	Prevented dopamine depletion and loss of Nissl and tyrosine hydroxylase positive neurons in substantia nigra.	(Matthews et al., 1999)
	Humans	Oral creatine, 20 g/d for 6 days, 2-4 g/d for 2 years	Reduced requirement for dose increases of dopaminergic therapy, and reported improved mental activity, behaviour and mood. No prevention of dopaminergic nerve cell loss.	(Bender et al., 2006)
Spinocerebell ar ataxia type I	Transgenic (B05) mouse	2% oral creatine initiated before onset of symptoms	Prevented the loss of Purkinje cells. No improvements in behavioural phenotype (rotarod, gait width test, open field test).	(Kaemmerer et al., 2001)

Disorder	Model/Species	Creatine Treatment	Neuroprotective Outcomes	Reference
Amyotrophic lateral sclerosis	Transgenic (G93A) mouse	1-2% oral creatine initiated before onset of symptoms	Delayed onset and severity of clinical symptoms, improved survival and rotarod performance. Decreased oxidative injury, prevented loss of substantia nigra and motor neurons, and prevented a decrease of cholinergic neuronal activity in the olfactory cortex and hippocampus, although not in lumbar spinal cord.	(Klivenyi et al., 1999, Pena-Altamira et al., 2005, Snow et al., 2003)
	Humans	Oral creatine ranging from 5 g/d for 9 months to 10 g/d for 1 year	Prevented the reduction of metabolite levels in motor cortex, but mixed results for improved survival (trend towards increase or no effect). No improvement shown in the rate of decline of motor, respiratory or functional capacity.	(Groeneveld et al., 2003, Rosenfeld et al., 2008, Shefner et al., 2004, Vielhaber et al., 2001)
Traumatic brain injury	Controlled cortical contusion in rats	1% oral creatine 2-4 weeks before and 1 week after injury	Reduced markers of secondary cellular injury in cortex, penumbra and hippocampus 6 h after injury, and cortical tissue sparing and reduced mitochondrial swelling and dysfunction up to 1 week after injury.	(Scheff and Dhillon, 2004, Sullivan et al., 2000)
	Humans – adults and children	0.4 g/kg/d oral creatine 6 months after injury	Reduced duration of intubation, time in intensive care and post-traumatic amnesia. In adults, improved cognition, personality, self care and communication at 6 months after injury. In children, reduced headache, dizziness and fatigue.	(Sakellaris et al., 2006, Sakellaris et al., 2008)

Table 1. (Continued)

Disorder	Model/Species	Creatine Treatment	Neuroprotective Outcomes	Reference
Spinal cord injury	Spinal cord contusion in rats	2-5% oral creatine 4-5 weeks before and after injury	5% dose showed reduced scar tissue and improved locomotor capacity 1-2 weeks after injury. 2% dose showed no beneficial effect on hindlimb function or white matter sparing. Some sparing of grey matter with post-injury creatine supplementation.	(Hausmann et al., 2002, Rabchevsky et al., 2003)
Adult stroke	Transient focal ischemia in rats	2% oral creatine 3-4 weeks before injury	Reduced infarct volume by 40-60%, enhanced vasodilator response, improved post-ischemic recovery of cerebral blood flow, and reduced markers of apoptosis and improved neurological function at 24 h. Inconsistent reports for preservation of creatine and ATP in post-ischemic brain.	(Prass et al., 2006, Zhu et al., 2004)
	Transient global ischemia in rats	50 mM creatine ICV infused over 5 days before and 7 days after injury	Prevented abnormal neuronal morphology, reduced glial response, and reduced pyknotic and apoptotic cell death in neocortex, hippocampus, and caudate putamen.	(Lensman et al., 2006, Otellin et al., 2003)
	Transient hypoxia in rats	1 mg/g creatine IP 30 min before injury	Prevented increase in marker of lipid peroxidation immediately after hypoxia.	(Rauchova et al., 2002)
	Rat hippocampal slices under anoxic conditions	1-10 mM creatine pre-treatment	Delayed anoxic depolarization and prevented irreversible loss of synaptic transmission. Reduced post-hypoxic hyperexcitability with higher doses (5-10 mM).	(Balestrino et al., 2002, Balestrino et al., 1999, Parodi et al., 2003, Zapara et al., 2004)

ATP, adenosine triphosphate; d, day; DNA, deoxyribonucleic acid; ICV, intracerebroventricular; IP, intraperitoneal; MPTP, 1-methyl-4-phenyl-1,2,3,6-tetrahydropyridine; 3-NP, 3-nitroproprionic acid.

However, there is currently no reason to think that the mechanisms of action of creatine – endogenous or supplemented – within the body are any different in carnivore and herbivore species.

Another consideration is the stage of development and maturation of the conceptus when birth normally occurs. Parturition occurs in most rodents when the fetuses are still relatively immature, and such altricial offspring undergo rapid maturation postnatally under maternal care. It is likely that intestinal uptake of creatine from milk provides for some of the creatine requirement until sufficient maturation of the renal-hepatic axis has occurred.

The relative inactivity of altricial, nest-bound neonates may have the effect of minimizing demand for creatine-phosphocreatine to support the maintenance of ATP in muscle, in particular. In contrast, where the fetus is born after a long gestation during which organ maturation and brain function is more advanced, the presence of a readily available creatine 'pool' may be more important, particularly when physical activity is required for survival. In such cases of precocial development it is to be expected that birth occurs when endogenous synthesis has also been attained. The human infant can be regarded as falling between these extremes, and as preterm birth is another characteristic of human pregnancy not shared by other species, it is likely that some preterm infants are unable to either synthesize sufficient creatine to meet demands, or absorb it from the gut (Lage et al, 2013).

Finally, the endocrinology of pregnancy is important because in some species, including the human, there is a complex interaction in the secretion and metabolism of steroid hormones that requires integration between the fetal adrenal gland, liver, and placenta. The presence of this 'feto-placental unit' has been documented for humans and sub-human primates (Ishimoto and Jaffe, 2011) and for the spiny mouse, a precocial rodent-like species (Quinn et al, 2013). While there is no evidence at present to suggest that creatine supplementation has any effect on the placental metabolism of fetal or maternal steroids, the impact of any maternal treatment in late pregnancy on this important system may need to be addressed.

Creatine and the Mother

Pregnancy is a period of increased nutritional requirements for the mother (Picciano, 2003), particularly in the latter part of gestation (Butte et al., 2004; Butte and King, 2005). As discussed below, maternal dietary creatine supplementation and fetal tissue loading during pregnancy has been explored as a treatment to protect the term fetus from episodes of oxygen deprivation during birth, but whether this could also improve fetal growth where placental insufficiency, infarction, or infection *in utero* is present warrants further investigation. The spiny mouse model of intrapartum birth asphyxia has shown that a range of benefits arise for the fetus/neonate when the maternal diet is supplemented with creatine from mid-gestation (Ireland et al., 2011, Ireland et al., 2008, Cannata et al., 2010, Ellery et al., 2012). While the ability of the creatine/ phosphocreatine system to maintain ATP turnover in cells when deprived of oxygen is likely to be the primary source of the benefit of creatine supplementation, other beneficial consequences are the continued shuttling of ATP from the mitochondria to areas of high-energy demand, and the antioxidant capacity attributed to creatine (Wallimann et al., 2011). However, the full effects of creatine dietary supplementation on the physiology of the female during pregnancy are yet to be examined. In non-pregnant adults the extended use of creatine supplementation for the purpose of improving exercise performance has shown occasional weight gain and changes in osmotic balance and blood pressure during high dose supplementation (Poortmans and Francaux, 2000). As pregnancy itself causes major changes in the set-points for regulating fluid balance and cardiorespiratory parameters (Guyton and Hall, 2006), there is a need to examine the effects of creatine on the physiology of the non-pregnant and pregnant female more closely before proceeding to clinical trials. There are also questions about the effect of chronic creatine supplementation on the *de novo* synthesis of creatine in the mother. It is to be expected that creatine supplementation will suppress AGAT expression, as shown for renal AGAT expression (Guthmiller et al., 1994, Wyss and Kaddurah-Daouk, 2000), but whether this occurs elsewhere (eg. pancreas, brain), and reverts fully after cessation of the treatment at parturition is yet to be determined.

Creatine and the Placenta

Where maternal creatine is passed to the fetal compartment, the placenta clearly has a critical role in the transport of creatine into fetal blood, but it may also retain some as a placental creatine 'pool'. We have recently provided the first report of mRNA expression of the creatine transporter (SCL6A8) in the two regions of the rodent placenta, suggesting transport of creatine to and from the fetal and maternal circulations (Dickinson et al., 2013). The concept of a placental pool of creatine from which fetal blood creatine is derived was first described by Miller in 1974 in the human placenta, and again in 1977 in the rat placenta, where bidirectional movement of creatine between the fetal to maternal circulations was demonstrated (Miller, 1974, Miller et al., 1977). Importantly though, the preferred direction of movement of creatine was from the mother to the fetus. A model for placental creatine transfer is that it is actively taken up into placental tissue from the maternal circulation by the creatine transporter (CrT), and then diffuses down a concentration gradient into the fetal circulation. Conceptually, this might mean that the release of creatine into the fetal circulation occurs as needed by the fetus for growth and its gradual increase in physical activity, and that this proceeds relatively separate from any fluctuations of creatine in maternal blood that might be affected by acute changes in the mother's activity, diet, or physical health.

Our recent work in the spiny mouse has shown significant loading of creatine in the placenta following maternal creatine supplementation (Ireland et al., 2008) suggesting, not withstanding the considerable amount of creatine that may be contained in blood in the vascular spaces of the placenta, that the placental creatine 'pool' can be readily increased. Studies in the spiny mouse also show that the placenta expresses relatively high levels of the CrT mRNA throughout pregnancy, and expression increases further just before birth (Ireland et al., 2009). This suggests that a significant part of embryonic/fetal creatine is derived from the mother via the placenta for most of pregnancy, and this may be critical for fetal development as data also suggests that the spiny mouse fetus is not capable of synthesizing its own creatine until after 0.9 day of gestation (Ireland et al., 2009).

The expression of creatine kinases (CK) in the rat placenta have been described and these are most highly expressed just prior to term (20 days of pregnancy) (Payne et al., 1993). The creatine kinases consist of a family of enzymes catalyzing the reversible transfer of high-energy phosphates from ATP to creatine, with subsequent generation of ADP and phosphocreatine. Two different isoenzymes of CK are necessary to establish an energy transfer system utilizing creatine as a phosphoryl carrier to regenerate ATP locally from ADP. In the rat uterus, expression of the 'brain' CK and ubiquitous mitochondrial CK mRNA and protein levels are coordinated, suggesting an important role for CK in the maintenance of pregnancy and labour.

It remains unclear if the placenta is able to synthesize creatine to any significant degree. The bulk of endogenously synthesized creatine arises from the production of guanidinoacetate (GAA) by activity of AGAT in the kidney, and the uptake of GAA by the liver where it is trans-methylated to form creatine by activity of guanidino-aminotransferase (GAMT) that is then released into the circulation. Minor systemic synthesis may occur in the pancreas, and the brain appears also to have independent synthesis to meets its own needs (Braissant et al, 2011).

The gene *GATM* that encodes AGAT, the rate-limiting step in the synthesis of creatine, has been identified in the mouse placenta where, furthermore, it appears to be a maternally imprinted gene (Sandell et al., 2003), although Ireland et al (2009) could not detect AGAT or GAMT mRNA or protein in spiny mouse placenta. The effect of chronic increases of maternal creatine during pregnancy on this aspect of placental metabolism therefore needs further examination. Maternally expressed genes in the placenta are thought to act to restrain the allocation of maternal resources to the embryo, acting to counteract paternally expressed genes that promote placental or embryonic growth.

A level of *GATM* expression in the placenta may provide for a means for the placenta to use the primary substrates for creatine synthesis (arginine, glycine, methionine) in the event that the fetal demand for creatine exceeds the maternal resource or capacity for the placenta to transfer creatine. This is a relatively unexplored finding, but suggests an important role for the creatine/phosphocreatine high-energy phosphate-buffering system during embryonic and fetal development.

Creatine and the Fetal Heart

The energy demand of the heart is high, with almost total reliance on oxidative pathways for ATP production and a very low capacity for anaerobic cellular metabolism. Heart muscle is therefore highly vulnerable to hypoxic injury. The creatine/phosphocreatine system plays an important role in the heart and serves to maintain ATP levels during hypoxic events (Ingwall and Weiss, 2004). Indeed, the myocardium has the second highest creatine pool in the body after skeletal muscle. In addition, we have reported an almost 15% increase in total creatine levels in the fetal heart after maternal creatine supplementation in the spiny mouse (Ireland et al. 2008), illustrating this organ's capacity for creatine loading even in fetal life.

In perinatology it is not surprising to find that the heart is affected in up to 30% of neonates following asphyxia at birth (Perlman et al., 1989, Martin-

Ancel et al., 1995), with transient myocardial ischemia leading to cardiac problems including left ventricular dysfunction, and ECG abnormalities. Increased myocardial creatine would therefore likely protect the myocardium during periods of ischemia and protect the muscle from damage. Our recent work with the spiny mouse model of birth asphyxia has shown that deleterious changes in ventricular function observed in animals asphyxiated at birth, including decreased left ventricular contractility and diastolic relaxation, are not observed in animals born to creatine-supplemented mothers (La Rosa et al, unpublished data). In support of these positive effects, a recent meta-analysis investigating the effect of creatine phosphate treatment given to asphyxiated newborns after diagnosis of myocardial damage found that it significantly reduced the serum levels of markers of cardiac damage, and it also improved recovery of these infants and reduced their period of hospitalisation (Miao et al., 2012). This finding is mirrored in the adult literature with various studies reporting marked improvements in patients with chronic heart failure (Ferraro, 1990, Grazioli et al. 1992, Scattolin et al. 1993).

These observations highlight the potential for creatine to not only protect the heart from hypoxic injury if given beforehand, but also to limit damage and aid repair when given after the occurrence of hypoxic-ischemic insult.

Creatine and the Fetal Kidney

Creatine supplementation regimens are particularly relevant to renal physiology, as the kidney is a highly perfused organ with high cellular energy demands. This, coupled with the low glycolytic capacity of the renal tubules, renders the kidney highly susceptible to hypoxic-ischemia and reperfusion injury (Bonventre and Venkatachalam., 1998).

Acute kidney injury (AKI) and renal failure in the neonate have long been associated with birth asphyxia (Stapleton et al., 1987). The incidence of renal injury following birth asphyxia has been reported to be 50% to 72% in various studies, with asphyxiated babies presenting with both oliguric and non-oliguric renal failure (Hankins et al., 2002, Martìn-Ancel et al., 1995, Jayashree et al., 1991). Whilst oliguria is more common, suggesting tubular deficiencies, the presentation of non-oliguric cases indicates hypoxic effects on glomerular filtration rate and possible glomerular damage (Aldana et al., 1995, Periman and Tack, 1988, Misra et al., 1991).

Studies also indicate that renal damage caused by birth asphyxia can lead to renal complications later in life - up to 40% of survivors of oliguric AKI show a decreased glomeruli filtration rate and develop tubular dysfunction leading to chronic kidney disease (Roberts et al., 1990). AKI in the neonatal period is also associated with an increased risk of hypertension later in life (Askenazi et al., 2009). Renal compromise can also arise because of the increased workload imposed on the kidneys by conditions such as acidosis and rhabdomyolysis, arising from birth asphyxia-related injury occurring elsewhere in the body (Bosch et al., 2009).

Renal failure due to birth asphyxia can arise from three major pathogenic mechanisms: 1) reduction in renal blood flow because of increased renal vascular resistance which, if prolonged, leads to severe damage to the renal parenchyma; 2) proximal tubular lipid peroxidation and cytotoxicity caused by acidemia and free radical damage (Zager and Foerder, 1992); and 3) AKI secondary to rhabdomyolysis and the release of large amounts of myoglobin into the bloodstream after skeletal muscle breakdown (Kojima et al., 1985). Bearing this in mind, it can be hypothesized that creatine loading of the fetal kidney via the maternal diet could have both direct benefits (protection of intra-renal metabolic status) and indirect benefits (prevention of the cytotoxic effects arising from other tissues).

The effects of asphyxia on the fetal kidney have been described in several animal models. In fetal sheep, Ikeda et al (2000), demonstrated varying degrees of tubular necrosis in severely asphyxiated lambs. They concluded that, next to histological changes in brain white matter, damage to the cortical tubules of the kidneys was one of the earliest asphyxia-induced changes to be observed in their fetal lambs. In the precocial spiny mouse, intrapartum birth asphyxia resulted in increased numbers of shrunken glomeruli, tubular dilatation, and disturbance to tubular architecture at 24 h of postnatal age (Ellery et al., 2012). Shrunken glomeruli have also been reported for prematurely delivered baboons (Sutherland et al., 2012). In each of these animal models, as for the spiny mouse (Dickinson et al., 2005) and the human infant, nephrogenesis is largely complete by the time of birth..

Given the evident vulnerability of the developing kidney to damage resulting from global fetal oxygen deprivation, the potential of creatine supplementation to prevent some or all of this damage was determined using the precocial spiny mouse (Ellery et al., 2012). The 2-fold increase in mRNA of Neutrophil gelatinase-associated lipocalin (*Ngal*) induced by birth asphyxia at term did not occur for fetuses from mothers that had been given 5% creatine in their diet from mid pregnancy (Ellery et al., 2012). *Ngal* is an early marker of kidney injury (Mori and Nakao, 2007), and this result suggests that creatine loading either prevented asphyxia-induced transcription of *Ngal*, or that the intra-renal injury was less severe and therefore did not prompt increased expression of this injury marker. While such findings support the use of maternal creatine supplementation to ameliorate kidney injury associated with birth asphyxia, the consequences of birth asphyxia and creatine treatment for renal function in the neonate and young adult animal need to be ascertained to conclusively demonstrate that pregnancy-related creatine treatment can also reduce the risk of developing the chronic kidney complications often associated with neonatal AKI later in life.

Creatine and Fetal Muscle

Skeletal muscle tissue has the highest resting levels of creatine of any tissue in the body. Individual muscle fibres, whether phenotypically fast or slow (based on their speed of contraction, oxidative or glycolytic mode of energy production) require ATP to contract (Armstrong & Phelps, 1984, Asmussen et al., 2003). Muscle fibres are extremely vulnerable to the ATP depletion that occurs during severe hypoxia-ischemia. In the term fetus of many species, one of the first physiological responses to hypoxia is muscle vasoconstriction which has the function of redistributing part of the cardiac output away from a 'non-essential' tissue to maintain adequate perfusion of the brain, heart and adrenal glands – the so-called 'brain sparing' response (Giussani et al. 1994). Oxidative muscle fibres then become energy depleted, and although different fibre types have varying capacities for anaerobic ATP production, the utilisation of glycolysis results in the accumulation of oxygen free radicals. Significant reductions in the capacity for force production (Caron et al., 2009), increase in proteolysis, and muscle fibre atrophy (Deldicque et al., 2007, Sacheck et al., 2007) can then result. As a striated muscle, the diaphragm of the fetus/neonate is also subject to these asphyxia-induced injury, thereby compromising the ability of the neonate to establish effective ventilation.

In the spiny mouse we observed a 14% and 27% increase in total creatine levels in the fetal diaphragm and gastrocnemius muscles respectively after maternal creatine supplementation (Cannata et al. 2010, unpublished data), illustrating this organ's capacity for creatine loading, even in the fetus. Birth asphyxia in the spiny mouse resulted in significant muscle atrophy and reduction in Ca^{2+}-dependent force production (Cannata et al., 2010). However, if the mother had received extra creatine during pregnancy these asphyxia-induced deficits in structure and function of the diaphragm were completely prevented (Cannata et al., 2010), probably because this maternal dietary treatment leads to significant loading of the fetal diaphragm with creatine. Similar observations of the effects of birth asphyxia and protection by creatine have been made for hindlimb muscles of this precocial rodent (La Rosa et al, unpublished data).

Creatine and the Fetal Brain

As shown in Table 1, studies in both experimental animal models and humans suggest that creatine can slow the progress of neurodegenerative disease in the adult brain, and ameliorate some of the consequences of acquired CNS injury. Importantly for the fetus and newborn, there is growing experimental evidence that creatine *pre-treatment* can protect the immature brain from hypoxia-ischemia injury.

The widespread expression of the CrT in the fetal rat brain implies the immature brain may have a greater capacity to take up creatine from the circulation than does the mature brain (Braissant et al., 2005). In the postnatal day 10 rat brain, a significant increase in the PCr-NTP (nucleoside triphosphate) ratio could be achieved by subcutaneous injections of creatine, but no such increase occurred at postnatal day 20 (Holtzman et al., 1998). We have shown in the spiny mouse that supplementing the maternal diet with creatine from mid-pregnancy results in a small but significant increase in fetal brain reserves of creatine (Ireland et al., 2008). While much is yet to be understood about creatine entry into the immature brain (recently reviewed by Braissant et al., 2011), the apparently greater ease of entry of creatine into the developing CNS is encouraging for the use of creatine to prevent hypoxia-related perinatal brain injury.

In vitro experimental evidence: Pre-treatment of near-term (0.9 gestation) fetal guinea pig hippocampal slices with 3 mM creatine for 2 h prior to 30-40 min of oxygen-glucose-deprivation delayed the inhibition of protein synthesis, and also improved the recovery of protein synthesis (Berger et al., 2004a). Although creatine did not prevent ATP depletion during the period of oxygen-glucose deprivation, creatine was considered neuroprotective as prolonged inhibition of protein synthesis is an early marker for neuronal cell damage

(Berger et al., 2004a). Similar protection was provided against glutamate toxicity in cultured neurons from embryonic rat hippocampus, when 0.1-2 mM creatine was given either 24 h before or 2 h after the addition of glutamate (0.5-1 mM) to the incubation medium (Brewer and Wallimann, 2000). Creatine treatment was also shown to prevent glutamate-induced dendritic pruning and ATP depletion, which are *in vitro* markers of glutamate toxicity (Brewer and Wallimann, 2000).

Wilken et al (Wilken et al., 1998) pre-treated neonatal mouse brainstem slices with creatine either by feeding the pregnant mother a diet supplemented with creatine (2g/kg/day), or by incubating the slices for 3 h with 10 mM creatine. Following 30 min anoxia, ATP levels in control brainstem slices were depleted by 44%. Both of these creatine pre-treatments (nutrition or incubation) completely prevented the ATP decrease. Interestingly, and unlike adult hippocampal slices where creatine pre-treatment delays anoxic depolarization (Balestrino et al., 2002, Balestrino et al., 1999), the neonatal, creatine-treated brainstem slices showed an enhanced initial response to hypoxia – increased burst duration and amplitude of hypoglossal nerve activity. Control slices showed a mean increase of just 14%, whereas for the creatine pre-treated slices the increase was 29%. These findings suggest that exogenous creatine might augment ventilation and improve oxygen delivery during and after anoxia, and not only prolong the survival of cerebral synaptic transmission (Wilken et al., 1998), but also preserve cardio-respiratory function in the neonate.

In vivo experimental evidence: As the mouse, rat and rabbit are developmentally immature at the time of birth (Clancy et al., 2001, Dobbing & Sands, 1979), hypoxia-ischemia is often induced in the postnatal period when the stage of brain development is more similar to the newborn human infant (Vannucci and Vannucci, 2005). Transient hypoxia-ischemia is induced by unilateral carotid artery ligation, followed by a period of breathing a hypoxic gas mixture (usually 4-8% oxygen, 92-96% nitrogen). In 10 day old rats, and 15 and 20 day old rabbit pups, pre-treatment with creatine (3 g/kg/day) for 3 days prior to insult has been shown to significantly reduce electroencephalographic and behavioural seizures during exposure to the hypoxic gas (Holtzman et al., 1998, Holtzman et al., 1999); creatine also significantly decreased the mortality rate immediately following the hypoxic insult (Holtzman et al., 1998). Creatine treated rat pups showed recovery of the PCr/NTP ratio within 2 h of re-breathing room air, whereas it fell significantly in saline-injected pups. A similar regimen of creatine pre-treatment followed by hypoxia-ischemia in 7 day old rats significantly reduced brain oedema when compared to saline-injected controls; indeed, the oedema was limited to the ipsilateral cortex, and unlike control rats, the hippocampus and basal ganglia were spared (Adcock et al., 2002b). Others have reported that pre- and post-treatment with creatine in 7 day old rat pups significantly reduced brain injury (Berger et al., 2004a). The difference between controls and treated animals 7 days after the insult was described as changing from severe cystic cerebral infarction to mild-moderate brain damage in treated animals, because the creatine treatment significantly reduced neuronal injury in the cortex and hippocampus, and prevented shrinkage of the ipsilateral hemisphere in comparison to the control hemisphere (Berger et al., 2004a). In a spiny mouse model of birth asphyxia (Ireland et al., 2008), creatine supplementation of the maternal diet for the last half of pregnancy almost completely prevented asphyxia-induced increase in expression of the pro-apoptotic protein Bax, and the increase of cytoplasmic cytochrome c protein in the cortical subplate, thalamus and piriform cortex, and lipid peroxidation in the cerebrum (Ireland et al., 2011). The neonates from the creatine-fed mothers also grew normally, whereas those from unsupplementated mothers showed postnatal growth retardation (Ireland et al., 2008).

Strengths of Creatine As a Protective Agent for the Developing Fetus and Newborn

The studies detailed in this review provide evidence that creatine is effective when used as a prophylactic treatment to protect the fetus and neonate from transient hypoxia. Unlike other neuroprotective therapies currently being evaluated (e.g., MgSO4), or which are entirely 'rescue' treatments (e.g., hypothermia), creatine has multiple mechanisms of action that lend it to protecting several major organs in the fetus and neonate, as well as the brain.

This topic has been dealt with in detail elsewhere (Wallimann et al., 2011, Dickinson et al., 2013), but the sum of these manifold actions is likely to be that creatine and PCr strengthen overall cellular energetics, prolonging the viability of tissues and providing them with resistance to the initial phase of a hypoxic-ischemic challenge.

Creatine also targets many of the secondary responses to hypoxic-ischemia. With creatine already in wide use in humans, and with few (if any) apparent side effects noted (for example, see: Kim et al, 2011; Gualano et al 2012), it has the potential to be easily incorporated into obstetric practice in a similar way that pre-conceptional folate is now routinely used to reduce the risk of neural tube defects. The relatively low-cost and ease of administration of creatine may be particularly relevant for developing countries where intrapartum asphyxia remains a leading cause of neonatal mortality (Azra Haider and Bhutta, 2006).

Conclusion

One of the greatest challenges in developing an effective treatment for perinatal hypoxia-ischemia is addressing the complex, damaging events that occur with global energy failure. The only specific treatment available at the present time is cooling of the head and total body (Shah et al., 2007). Although this treatment is effective, the therapeutic window is short (perhaps 2-6 h after birth) and it protects against only the later phase of secondary energy depletion; further, it has proven to have a much smaller effect in humans than in animal models (Graham et al., 2008).

The present review details significant evidence that creatine is a clinically relevant and promising prophylactic treatment to protect the fetus and neonate from transient hypoxia. Future research should be directed at confirming these findings in larger animal models of hypoxia-ischemia, before optimizing a safe and efficient dose for recommendation to pregnant women as an adjunctive therapy to help prevent neonatal brain injury.

All authors have no financial disclosures to make, and do not have any knowledge of conflicts of interest.

References

Adcock, K. H., Nedelcu, J., Loenneker, T., Martin, E., Wallimann, T., and Wagner, B. P. 2002a. Neuroprotection of creatine supplementation in neonatal rats with transient cerebral hypoxia-ischemia. *Developmental Neuroscience,* 24, 382-8.

Adcock, K. H., Nedelcu, J., Loenneker, T., Martin, E., Wallimann, T., and Wagner, B. P. 2002b. Neuroprotection of creatine supplementation in neonatal rats with transient cerebral hypoxia-ischemia. *Developmental Neuroscience*, 24, 382-8.

Aldana, V. C., Romaro, M. S., Vargas, O. A., and Hernandez, A. J. 1995. Acute complications in full term neonates with severe neonatal asphyxia. *Ginecologica y obstetricia de Mexico*, 63, 123.

Armstrong, R. B. and Phelps, R. O. 1984. Muscle fiber type composition of the rat hindlimb. *Am. J. Anat.*, 171, 259-72.

Askenazi, D., Ambalavanan, N. and Goldstein, S. 2009. Acute kidney injury in critically ill newborns: What do we know? What do we need to learn? *Pediatric Nephrology*, 24, 265-274.

Asmussen, G., Schmalbruch, I., Soukup, T., and Pette, D. 2003. Contractile properties, fiber types, and myosin isoforms in fast and slow muscles of hyperactive Japanese waltzing mice. *Exp. Neurol.*, 184, 758-66.

Azra Haider, B. and Bhutta, Z. A. 2006. Birth asphyxia in developing countries: current status and public health implications. *Current Problems in Pediatric and Adolescent Health Care*, 36, 178-88.

Balestrino, M., Lensman, M., Parodi, M., Perasso, L., Rebaudo, R., Melani, R., Polenov, S., and Cupello, A. 2002. Role of creatine and phospho-creatine in neuronal protection from anoxic and ischemic damage. *Amino Acids*, 23, 221-9.

Balestrino, M., Rebaudo, R. and Lunardi, G. 1999. Exogenous creatine delays anoxic depolarization and protects from hypoxic damage: dose-effect relationship. *Brain Research*, 816, 124-30.

Bender, A., Auer, D. P., Merl, T., Reilmann, R., Saemann, P., Yassouridis, A., Bender, J., Weindl, A., Dose, M., Gasser, T., and Klopstock, T. 2005. Creatine supplementation lowers brain glutamate levels in Huntington's disease. *Journal of Neurology*, 252, 36-41.

Bender, A., Koch, W., Elstner, M., Schombacher, Y., Bender, J., Moeschl, M., Gekeler, F., Muller-Myhsok, B., Gasser, T., Tatsch, K., and Klopstock, T. 2006. Creatine supplementation in Parkinson disease: a placebo-controlled randomized pilot trial. *Neurology*, 67, 1262-4.

Berger, R., Middelanis, J., Vaihinger, H.-M., Mies, G., Wilken, B., and Jensen, A. 2004a. Creatine protects the immature brain from hypoxic-ischemic injury. *Journal of the Society for Gynaecologic Investigation*, 11, 9-15.

Berger, R., Middelanis, J., Vaihinger, H. M., Mies, G., Wilken, B., and Jensen, A. 2004b. Creatine protects the immature brain from hypoxic-ischemic injury. *Journal of the Society for Gynecologic Investigation*, 11, 9-15.

Bonventre, J. V., M. Brezis, N. Siegel, S. Rosen, D. Portilla, and Venkatachalam, M. 1998. Acute renal failure. I. Relative importance of proximal vs. distal tubular injury. *American Journal of Physiology - Renal Physiology,* 275, F623-F632.

Bosch, X., Poch, E. and Grau, J. 2009. Rhabdomyolysis and acute kidney injury. *New England Journal of Medicine,* 361, 62.

Braissant, O., Beard, E. and Uldry, J. 2011. Models of creatine deficiency syndromes by RNAi in 3D reaggregated brain cell organotypic cultures. *15th International Symposium on Cerebral Blood Flow, Metabolism, and Function.* Barcelona, Spain.

Braissant, O., Henry, H., Villard, A.-M., Speer, O., Wallimann, T., and Bachmann, C. 2005. Creatine synthesis and transport during rat embryogenesis: spatiotemporal expression of AGAT, GAMT and CT1. *BMC Developmental Biology,* 5, 9.

Brewer, G. J. and Wallimann, T. W. 2000. Protective effect of the energy precursor creatine against toxicity of glutamate and beta-amyloid in rat hippocampal neurons. *Journal of Neurochemistry,* 74, 1968-1978.

Butte, N. F. and King, J. C. 2005. Energy requirements during pregnancy and lactation. *Public Health Nutrition-Cab International-,* 8, 1010.

Butte, N. F., Wong, W. W., Treuth, M. S., Ellis, K. J., and Smith, E. O. Ä. 2004. Energy requirements during pregnancy based on total energy expenditure and energy deposition. *The American journal of clinical nutrition,* 79, 1078-1087.

Calvert, J. W. and Zhang, J. H. 2005. Pathophysiology of an hypoxic-ischemic insult during the perinatal period. *Neurological Research,* 27, 246-60.

Cannata, D. J., Ireland, Z., Dickinson, H., Snow, R. J., Russell, A. P., West, J. M., and Walker, D. W. 2010. Maternal creatine supplementation from mid-pregnancy protects the diaphragm of the newborn spiny mouse from intrapartum hypoxia-induced damage. *Pediatr. Res.,* 68, 393-8.

Caron, A. Z., Drouin, G., Desrosiers, J., Trensz, F., and Grenier, G. 2009. A novel hindlimb immobilization procedure for studying skeletal muscle atrophy and recovery in mouse. *J. Appl. Physiol.,* 106, 2049-59.

Clancy, B., Darlington, R. B. and Finlay, B. L. 2001. Translating developmental time across mammalian species. *Neuroscience,* 105, 7-17.

Dedeoglu, A., Kubilus, J. K., Yang, L., Ferrante, K. L., Hersch, S. M., Beal, M. F., and Ferrante, R. J. 2003. Creatine therapy provides neuroprotection after onset of clinical symptoms in Huntington's disease transgenic mice. *Journal of Neurochemistry,* 85, 1359-67.

Deldicque, L., Theisen, D., Bertrand, L., Hespel, P., Hue, L., and Francaux, M. 2007. Creatine enhances differentiation of myogenic C2C12 cells by activating both p38 and Akt/PKB pathways. *Am. J. Physiol. Cell Physiol.,* 293, C1263-71.

Dickinson, H., Ireland, Z. J., Larosa, D. L., O'Connell, B., Ellery, S. J., Snow, R., and Walker, D. W. 2013. Maternal dietary creatine supplementation does not alter the capacity for creatine synthesis in the newborn spiny mouse. *Reproductive sciences,* In Press.

Dickinson, H., Walker, D. W., Cullen-Mcewen, L., Wintour, E. M., and Moritz, K. 2005. The spiny mouse (*Acomys cahirinus*) completes nephrogenesis before birth. *American Journal of Physiology - Renal Physiology,* 289, F273-9.

Dobbing, J. and Sands, J. 1979. Comparative aspects of the brain growth spurt. *Early Human Development,* 3, 79-83.

Ellery, S. J., Ireland, Z., Kett, M. M., Snow, R., Walker, D. W., and Dickinson, H. 2012. Creatine pretreatment prevents birth asphyxia-induced injury of the newborn spiny mouse kidney. *Pediatr. Res.,* epub. ahead of print.

Ferrante, R. J., Andreassen, O. A., Jenkins, B. G., Dedeoglu, A., Kuemmerle, S., Kubilus, J. K., Kaddurah-Daouk, R., Hersch, S. M., and Beal, M. F. 2000. Neuroprotective effects of creatine in a transgenic mouse model of Huntington's disease. *The Journal of Neuroscience,* 20, 4389-4397.

Ferraro, S. 1990. Acute and short-term efficacy of high doses of creatine phosphate in the treatment of cardiac failure. *Current Therapeutic Research, Clinical and Experimental,* 47, 917.

Giussani, D., Spencer, J. A. D., Hanson, M. A. 1994. Fetal cardiovascular responses to hypoxaemia. *Fetal Maternal Med. Rev.,* 6: 17-37.

Grazioloi, I., Melzi, G. and Strumia, E., 1992. Multicenter controlled study of creatine phosphate in the treatment of heart failure. *Current Therapeutic Research* 52: 271-280.

Graham, E. M., Ruis, K. A., Hartman, A. L., Northington, F. J., and Fox, H. E. 2008. A systematic review of the role of intrapartum hypoxia-ischemia in the causation of neonatal encephalopathy. *American Journal of Obstetrics and Gynecology,* 199, 587-595.

Groeneveld, G. J., Veldink, J. H., Van Der Tweel, I., Kalmijn, S., Beijer, C., De Visser, M., Wokke, J. H. J., Franssen, H., and Van Den Berg, L. H. 2003. A randomized sequential trial of creatine in amyotrophic lateral sclerosis. *Annals of Neurology,* 53, 437-45.

Gualano B., Roschel H., Lancha-Jr A.H., Brightbill C.E., Rawson E.S. 2012. In sickness and in health: the widespread application of creatine supplementation. *Amino Acids,* 43, 519-29.

Guthmiller, P., Van Pilsum, J. F., Boen, J. R., and McGuire, D. M. 1994. Cloning and sequencing of rat kidney L-arginine:glycine amidinotransferase. Studies on the mechanism of regulation by growth hormone and creatine. *Journal of Biological Chemistry,* 269, 17556-60.

Guyton, A., Hall, J. E. 2006, *Textbook of Medical Physiology* (11 ed.). Philadelphia: Saunders. pp. 103

Hankins, G. D. V., Koen, S., Gei, A. F., Lopez, S. M., Van Hook, J. W., and Anderson, G. D. 2002. Neonatal organ system injury in acute birth asphyxia sufficient to result in neonatal encephalopathy. *Obstetrics and Gynecology,* 99, 688.

Hausmann, O. N., Fouad, K., Wallimann, T., Schwab, M. E., and Klasman, I. 2002. Protective effects of oral creatine supplementation on spinal cord injury in rats. *Spinal Cord,* 40, 449-456.

Hersch, S., Gevorkian, S., Marder, K., Moskowitz, C., Feigin, A., Cox, M., Como, P., Zimmerman, C., Lin, M., Zhang, L., Ulug, A., Beal, M., Matson, W., Bogdanov, M., Ebbel, E., Zaleta, A., Kaneko, Y., Jenkins, B., hevelone, N., Zhang, H., Yu, H., Schoenfeld, D., Ferrante, R., and Rosas, H. 2006. Creatine in Huntington disease is safe, tolerable, bioavailable in brain and reduces serum 8OH2'dG. *Neurology,* 66, 250-2.

Holtzman, D., Khait, I., Mulkern, R., Allred, E., Rand, T., Jensen, F., and Kraft, R. 1999. In vivo development of brain phosphocreatine in normal and creatine-treated rabbit pups. *Journal of Neurochemistry,* 73, 2477-2484.

Holtzman, D., TogliattI, A., Khait, I., and Jensen, F. 1998. Creatine increases survival and suppresses seizures in the hypoxic immature rat. *Paediatric Research,* 44, 410-414.

Ikeda, T., Murata, Y., Quilligan, E. J., Parer, J. T., Murayama, T., and Koono, M. 2000. Histologic and biochemical study of the brain, heart, kidney, and liver in asphyxia caused by occlusion of the umbilical cord in near-term fetal lambs. *American journal of obstetrics and gynecology,* 182, 449-457.

Ingwall, J. S. and Weiss, R. G. 2004. Is the failing heart energy starved? on using chemical energy to support cardiac function. *Circ. Res.,* 95, 135-45.

Ireland, Z., Castillo-Melendez, M., Dickinson, H., Snow, R., and Walker, D. W. 2011. A maternal diet supplemented with creatine from mid-pregnancy protects the newborn spiny mouse brain from birth hypoxia. *Neuroscience,* 194, 372-9.

Ireland, Z., Dickinson, H., Snow, R., and Walker, D. W. 2008. Maternal creatine: does it reach the fetus and improve survival after an acute hypoxic episode in the spiny mouse (Acomys cahirinus)? *Am. J. Obstet. Gynecol.,* 198, 431 e1-6.

Ireland, Z., Russell, A. P., Wallimann, T., Walker, D. W., and Snow, R. 2009. Developmental changes in the expression of creatine synthesizing enzymes and creatine transporter in a precocial rodent, the spiny mouse. *BMC Dev. Biol.,* 9, 39.

Ishimoto, H. and Jaffe, R. B. 2011. Development and function of the human fetal adrenal cortex: a key component in the feto-placental unit. *Endocr. Rev.,* 32, 317-55.

Jayashree, G., Dutta, A., Sarna, M., and Saili, A. 1991. Acute renal failure in asphyxiated newborns. *Indian pediatrics,* 28, 19-23.

Kaemmerer, W. F., Rodrigues, C. M. P., Steer, C. J., and Low, W. C. 2001. Creatine-supplemented diet extends Purkinje cell survival in spinocerebellar ataxia type 1 transgenic mice but does not prevent the ataxic phenotype. *Neuroscience,* 103, 713-724.

Kim H.J., Kim C.K., Carpentier A., Poortmans J.R., 2011. Studies on the safety of creatine supplementation. *Amino Acids,* 40, 1409-18.

Klivenyi, P., Ferrante, R., Matthews, R., Bogdanov, M., Klein, A., and Facorro, G. 1999. Neuroprotective effects of creatine in a transgenic animal model of amyotrophic lateral sclerosis. *Nature Medicine,* 5, 347-50.

Kojima, T., Kobayashi, T., Matsuzaki, S., Iwase, S., and Kobayashi, Y. 1985. Effects of perinatal asphyxia and myoglobinuria on development of acute, neonatal renal failure. *Archives of Disease in Childhood,* 60, 908-912.

Lage, S., Andrade, F., Prieto, J. A., Asla, I., Rodriguez, A., Ruiz, N., Echeverria, J. Couce, M. L., Sanjurjo, P., and Aldamiz-Echevarria, L. 2013. Arginine-guanidinoacetate-creatine pathway in preterm newborns: creatine biosynthesis in newborns. *J. Pediatr. Endocr. Metab.,* 26(1-2), 53-60.

Lensman, M., Korzhevskii, D. E., Mourovets, V. O., Kostkin, V. B., Izvarina, N., Perasso, L., Gandolfo, C., Otellin, V. A., Polenov, S. A., and Balestrino, M. 2006. Intracerebroventricular administration of creatine protects against damage by global cerebral ischemia in rat. *Brain Research,* 1114, 187-194.

Martìn-Ancel, A., Garcìa-Alix, A., Cabanas, F. G. F., Burgueros, M., and Quero, J. 1995. Multiple organ involvement in perinatal asphyxia. *The Journal of Pediatrics,* 127, 786-793.

Martin-Ancel, A., Garcia-Alix, A., Gaya, F., Cabanas, F., Burgueros, M., and Quero, J. 1995. Multiple organ involvement in perinatal asphyxia. *J. Pediatr.*, 127, 786-93.

Matthews, R. T., Ferrante, R. J., Klivenyi, P., Yang, L., Klein, A. M., Mueller, G., Kaddurah-Daouk, R., and Beal, M. F. 1999. Creatine and cyclocreatine attenuate MPTP neurotoxicity. *Experimental Neurology*, 157, 142-9.

Matthews, R. T., Yang, L., Jenkins, B. G., Ferrante, R. J., Rosen, B. R., Kaddurah-Daouk, R., and Beal, M. F. 1998. Neuroprotective effects of creatine and cyclocreatine in animal models of Huntington's disease. *The Journal of Neuroscience*, 18, 156-63.

Miao, P., Sun, B. and Feng, X. 2012. [Treatment of myocardial damage with creatine phosphate following neonatal asphyxia: a meta-analysis]. *Zhongguo Dang Dai Er Ke Za Zhi*, 14, 172-6.

Miller, R. 1974. Transport of creatine in the human placenta. *The Pharmacologist*, 16.

Miller, R. K., Davis, B. M., Brent, R. L., and Koszalka, T. R. 1977. Creatine transport by rat placentas. *Am. J. Physiol.*, 233, E308-15.

Misra, P., Kumar, A., Natu, S., Kapoor, R., Srivastava, K., and Das, K. 1991. Renal failure in symptomatic perinatal asphyxia. *Indian pediatrics*, 28, 1147-1151.

Mori, K. and Nakao, K. 2007. Neutrophil gelatinase-associated lipocalin as the real-time indicator of active kidney damage. *Kidney international*, 71, 967-970.

Nordstrom, L. and Arulkumaran, S. 1998. Intrapartum fetal hypoxia and biochemical markers: a review. *Obstet. Gynecol. Surv.*, 53, 645-57.

Ohtsuki, S., Tachikawa, M., Takanaga, H., Shimizu, H., Watanabe, M., Hosoya, K.-I., and Terasaki, T. 2002. The blood-brain barrier creatine transporter is a major pathway for supplying creatine to the brain. *Journal of Cerebral Blood Flow and Metabolism*, 22, 1327-35.

Otellin, V., Korzhevskii, D., Kostkin, V., Balestrino, M., Lensman, M., and Polenov, S. 2003. The neuroprotective effect of creatine in rats with cerebral ischemia. *Doklady Biological Sciences*, 390, 197-199.

Parodi, M., Rebaudo, R., Perasso, L., Gandolfo, C., Cupello, A., and Balestrino, M. 2003. Effects of exogenous creatine on population spike amplitude and on postanoxic hyperexcitability in brain slices. *Brain Research*, 963, 197-202.

Payne, R. M., Friedman, D. L., Grant, J. W., Perryman, M. B., and strauss, A. W. 1993. Creatine kinase isoenzymes are highly regulated during pregnancy in rat uterus and placenta. *Am. J. Physiol.*, 265, E624-35.

Pena-Altamira, E., Crochemore, C., Virgili, M., and Contestabile, A. 2005. Neurochemical correlates of differential neuroprotection by long-term dietary creatine supplementation. *Brain Research,* 1058, 183-188.

Periman, J. M. and Tack, E. D. 1988. Renal injury in the asphyxiated newborn infant: Relationship to neurologic outcome+. *The Journal of Pediatrics,* 113, 875-879.

Perlman, J. M., Tack, E. D., Martin, T., Shackelford, G., and Amon, E. 1989. Acute systemic organ injury in term infants after asphyxia. *Am. J. Dis. Child,* 143, 617-20.

Picciano, M. F. 2003. Pregnancy and lactation: physiological adjustments, nutritional requirements and the role of dietary supplements. *The Journal of Nutrition,* 133, 1997S-2002S.

Poortmans, J. R. and Francaux, M. 2000. Adverse effects of creatine supplementation: fact or fiction? / Effets secondaires nocifs de la supplementation en creatine: realite ou fiction? *Sports Medicine,* 30, 155-170.

Prass, K., Royl, G., Ute, L., Freyer, D., Megow, D., Dirnagl, U., Stockler-Ipsiroglu, G., Wallimann, T., and Priller, J. 2006. Improved reperfusion and neuroprotection by creatine in a mouse model of stroke. *Journal of Cerebral Blood Flow and Metabolism,* 27, 452-459.

Rabchevsky, A. G., Sullivan, P. G., Fugaccia, I., and Scheff, S. W. 2003. Creatine diet supplement for spinal cord injury: Influences on functional recovery and tissue sparing in rats. *Journal of Neurotrauma,* 20, 659-669.

Rauchova, H., Koudelova, J., Drahota, Z., and Mourek, J. 2002. Hypoxia-induced lipid peroxidation in rat brain and protective effect of carnitine and phosphocreatine. *Neurochemical Research,* 27, 899-904.

Roberts, D. S., Haycock, G. B., Dalton, R. N., Turner, C., Tomlinson, P., Stimmler, L., and Scopes, J. W. 1990. Prediction of acute renal failure after birth asphyxia. *Archives of Disease in Childhood,* 65, 1021-1028.

Rosenfeld, J., King, R. M., Jackson, C. E., Bedlack, R. S., Barohn, R. J., Dick, A., Phillips, L. H., Chapin, J., Gelinas, D. F., and Lou, J.-S. 2008. Creatine monohydrate in ALS: effects on strength, fatigue, respiratory status and ALSFRS. *Amyotrophic Lateral Sclerosis,* 9, 266-72.

Sacheck, J. M., Hyatt, J. P., Raffaello, A., Jagoe, R. T., Roy, R. R., Edgerton, V. R., Lecker, S. H., and Goldberg, A. L. 2007. Rapid disuse and denervation atrophy involve transcriptional changes similar to those of muscle wasting during systemic diseases. *FASEB J., 21,* 140-55.

Sakellaris, G., Kotsiou, M., Tamiolaki, M., Kalostos, G., Tsapaki, E., Spanaki, M., Spilioti, M., Charissis, G., and Evangeliou, A. 2006. Prevention of complications related to traumatic brain injury in children and adolescents with creatine administration: An open label randomized pilot study. *Journal of Trauma - Injury, Infection and Critical Care,* 61, 322-329.

Sakellaris, G., Nasis, G., Kotsiou, M., Tamiolaki, M., Charissis, G., and Evangeliou, A. 2008. Prevention of traumatic headache, dizziness and fatigue with creatine administration. A pilot study. *Acta pædiatrica,* 97, 31-4.

Sandell, L. L., Guan, X. J., Ingram, R., and Tilghman, S. M. 2003. Gatm, a creatine synthesis enzyme, is imprinted in mouse placenta. *Proc. Natl. Acad. Sci. US,* 100, 4622-7.

Scattolin, G., Gabellini, A., Desideri, M., Formidini, F., Caneve, F., Corbara, F. et al., 1993. Diastolic function and creatine phosphate: An echocardiographic study. *Current Therapeutic Research* 54: 562-571.

Scheff, S. W. and Dhillon, H. S. 2004. Creatine-enhanced diet alters levels of lactate and free fatty acids after experimental brain injury. *Neurochemical Research,* 29, 469-479.

Schulze, A. 2003. Creatine deficiency syndromes. *Molecular and Cellular Biochemistry,* 244, 143-50.

Shah, P. S., Ohlsson, A. and Perlman, M. 2007. Hypothermia to treat neonatal hypoxic ischemic encephalopathy: Systematic review. *Archives of Pediatrics and Adolescents Medicine,* 161, 951-958.

Shear, D. A., Haik, K. L. and Dunbar, G. L. 2000. Creatine reduces 3-nitropropionic-acid-induced cognitive and motor abnormalities in rats. *Neuroreport,* 11, 1833-7.

Shefner, J., Cudkowicz, M., Schoenfeld, D., Conrad, T., Taft, J., Chilton, M., Urbinelli, L., Qureshi, M., Zhang, H., Pestronk, A., Caress, J., Donofrio, P., Sorenson, E., Bradley, W., Lomen-Hoerth, C., Pioro, E., Rezania, K., Ross, M., Pascuzzi, R., Heiman-Patterson, T., Tandan, R., Mitsumoto, H., Rothstein, J., Smith-Palmer, T., Macdonald, D., and Burke, D. 2004. A clinical trial of creatine in ALS. *Neurology,* 63, 1656-61.

Snow, R. J., Turnbull, J., Da Silva, S., Jiang, F., and Tarnopolsky, M. A. 2003. Creatine supplementation and riluzole treatment provide similar beneficial effects in copper, zinc superoxide dismutase (G93A) transgenic mice. *Neuroscience,* 119, 661-7.

Stapleton, F. B., Jones, D. P. and Green, R. S. 1987. Acute renal failure in neonates: incidence, etiology and outcome. *Pediatr. Nephrol.,* 1, 314-20.

Sullivan, P. G., Geiger, J. D., Mattson, M. P., and Scheff, S. W. 2000. Dietary supplement creatine protects against traumatic brain injury. *Annals of Neurology,* 48, 723-729.

Sutherland, M. R., Yoder, B. A., McCurnin, D., Seidner, S., Gubhaju, L., Clyman, R. I., and Black, M. J. 2012. Effects of Ibuprofen treatment on the developing preterm baboon kidney. *American Journal of Physiology-Renal Physiology,* 302, 1286-1293.

Tabrizi, S. J., Blamire, A. M., Manners, D. N., Rajagopalan, B., Styles, P., Schapira, A. H. V., and Warner, T. T. 2005. High-dose creatine therapy for Huntington disease: a 2-year clinical and MRS study. *Neurology,* 64, 1655-6.

Vannucci, R. C. and Vannucci, S. J. 2005. Perinatal hypoxic-ischemic brain damage: evolution of an animal model. *Developmental Neuroscience,* 27, 81-6.

Vielhaber, S., Kaufmann, J., Kanowski, M., Sailer, M., Feistner, H., Tempelmann, C., Elger, C. E., Heinze, H. J., and Kunz, W. S. 2001. Effect of creatine supplementation on metabolite levels in ALS motor cortices. *Experimental Neurology,* 172, 377-82.

Wallimann, T., Tokarska-Schlattner, M. and Schlattner, U. 2011. The creatine kinase system and pleiotropic effects of creatine. *Amino Acids,* 40, 1271-96.

Wilken, B., Ramirez, J., Probst, I., Richter, D., and Hanefeld, F. 1998. Creatine protects the central respiratory network of mammals under anoxic conditions. *Paediatric Research,* 43, 8-14.

Wyss, M. and Kaddurah-Daouk, R. 2000. Creatine and creatinine metabolism. *Physiological Reviews,* 80, 1107-213.

Wyss, M. and Schulze, A. 2002. Health implications of creatine: can oral creatine supplementation protect against neurological and atherosclerotic disease? *Neuroscience,* 112, 243-60.

Zager, R. and Foerder, C. 1992. Effects of inorganic iron and myoglobin on in vitro proximal tubular lipid peroxidation and cytotoxicity. *Journal of Clinical Investigation,* 89, 989.C

Zapara, T. A., Simonova, O. G., Zharkikh, A. A., Balestrino, M., and Ratushniak, A. S. 2004. Seasonal differences and protection by creatine or arginine pretreatment in ischemia of mammalian and molluscan neurons in vitro. *Brain Research,* 1015, 41-9.

Zhu, S., Li, M., Figueroa, B. E., Liu, A., Stavrovskaya, I. G., Pasinelli, P., Beal, M. F., Brown, R. H., Jr, Kristal, B. S., Ferrante, R. J., and Friedlander, R. M. 2004. Prophylactic creatine administration mediates neuroprotection in cerebral ischemia in mice. *The Journal of Neuroscience, 24,* 5909-5912.

In: Creatine
Editors: F. D'Cruz and V. Ribeiro

ISBN: 978-1-62948-295-8
© 2013 Nova Science Publishers, Inc.

Chapter V

Creatine: Metabolism and Role in Sports Physiology

*Ayed Ikram Bezrati-Ben[1], Fehmi Nasrallah[2],
Raouf Hammami[3], Karim Chamari[4], Moncef Feki[5]
and Naziha Kaabachi[5]**

[1]Laboratory of Biochemistry, Rabta Hospital, Tunisia;
National Centers for Medicine and Sciences of Sports;
Faculty of Medicine of Tunis, El Manar University, Tunis, Tunisia
[2]Laboratory of Biochemistry, Rabta Hospital, Tunisia
[3]Research Laboratory "Sport Performance Optimization''National
Centers for Medicine and Sciences of Sports, Tunisia
[4]Research & Education Center Aspetar, Doha, Qatar
[5]Laboratory of Biochemistry, Rabta Hospital, Tunisia;
Faculty of Medicine of Tunis, El Manar University, Tunis, Tunisia

---

* Corresponding author: Naziha Kaabachi, Laboratory of Biochemistry, Rabta Hospital, 1007
Jebbari, Tunis, Tunisia. Tel/fax, 00 216 71 561 912; E-mail address:
Naziha.kaabachi@rns.tn; Naziha.kaabachi@gmail.com

Abstract

Methyl guanidine-acetic acid, commonly named creatine (Cr), is a naturally occurring amino acid provided by diet or synthesized endogenously, primarily in liver, pancreas and kidneys. Cr biosynthesis involves two enzymes; arginine glycine amidinotransferase that converts arginine and glycine to ornithine and guanidinoacetate, and S-adenosyl-L-methionine:N-guanidinoacetate methyltransferase that converts guanidinoacetate to Cr. Cr is taken up from blood into Cr-requiring tissues against a concentration gradient by a specific transporter (SLC6A8). Approximately 95% of Cr is stored in skeletal muscle and the remaining 5% is stored in the heart, brain and testes. Cr and its derivative phosphocreatine (PCr) play an essential role in energy storage and transmission in most tissues; predominant role being played in skeletal muscle and brain. In athletes, PCr is particularly important during explosive activities when a high rate of energy release is required. PCr serves as a source of high energy phosphates and increasing PCr muscle stores might have a significant effect on physical performance, especially during anaerobic intense exercises. Several researchers have evaluated the potential ergrogenic value of Cr. It has been shown that Cr supplementation increases total Cr content in skeletal muscle, particularly in individuals with lowest basal Cr content. Numerous studies have examined the effect of Cr supplementation in athletes. Most studies have shown that Cr supplementation is accompanied by a significant increase in body weight, with increase of lean mass and decrease of fat mass. This may be due to stimulation of muscle protein synthesis or water retention, a point that is particularly relevant for athletes in explosive sports disciplines. The results support the proposal that Cr supplementation improves performance particularly during high intensity exercise, as well as resistance training. However, other studies failed to show a significant effect of Cr supplementation. Issues such as the dose and duration of supplementation, exercise protocol, and sports disciplines may have contributed to these discrepancies. There are some anecdotal reports on potentially adverse effects of Cr supplementation such as muscle cramps, gastrointestinal complaints, liver dysfunction, and kidney impairment. It appears that Cr supplementation at the correct dose is not harmful to healthy athletes, but it must be used with precaution in individuals at risk. Finally, it seems reasonable to monitor liver and kidney functions in individuals consuming Cr supplements.

Keywords: Athlete, creatine, exercise, physical performance

Historical Background

Creatine (Cr) was initially identified by the French scientific Chevreul in the early 1830s, when he confirmed it as an organic constituent of meat. In 1847, Leibeg observed that the meat of wild foxes killed in the chase contained 10 times Cr as that of foxes in captivity. He thus concluded that Cr was somehow linked to muscle performance.

In 1912, Denis and Folin reported that after ingestion of Cr, cat muscle Cr stores increased by 70%. In 1927 and 1929, Fiske and Subbarow identified phosphocreatine or creatine phosphate (PCr) in the resting muscle of cat [1]. They reported then that during electrical stimulation of skeletal muscle, PCr decreased but then increased to initial levels after a recovery period. However, it was not until the beginning of the 20th century that Cr ingestion research was developed. Since, investigators began to utilize Cr as a potential factor for performance improvement [2-6].

Metabolism of Creatine

Cr or Methyl guanidine-acetic acid is a naturally occurring amino acid provided by exogenous sources such as fish and meat and also can be synthesized endogenously in internal organs, predominately in the liver at an amount of about 1 g per day [7, 8]. The synthesis of Cr involves the 3 aminoacids glycine, arginine, and methionine. Initially, arginine and glycine combine to form guanidinoacetate and ornithine, a reversible reaction catalyzed by the enzyme glycine amidine transamidinase (AGAT).

Then Cr is formed by the addition of methyl group from S-adenosylmethionime involving the enzyme methyltransferase (GAMT) for the irreversible reaction [7]. Cr is synthesized outside of muscles and thus must be transported to skeletal muscle and requiring tissues via the blood stream. Cr enters the muscle cell against a concentration gradient with the aid of a sodium chloride dependent Cr transporter [9].

The majority of Cr in the human body is stored in skeletal muscle in two forms, approximately 60% is stored in the phosphorylated form and 40% in the free from.

An average young male rocighing 70 Kg has a Cr pool of approximately 120 to 140 g [10, 11], which varies between individuals depending on muscle mass and skeletal muscle fiber type. However, rate of PCr is correlated with the glycolytic capacity of the skeletal muscle fiber type. As PCr concentration in type II muscle fibers is significantly higher than type I muscle fiber, the degradation rate of PCr is faster in the type II muscle fiber, but their regeneration is faster in type I muscle fiber [12, 13].

Regulation of Creatine Metabolism

Cr uptake is regulated by various mechanisms, mainly changes in expression levels of the enzymes involved in Cr metabolism as well as extracellular levels of Cr. In addition, changes in the transport capacity and permeability of biological membrane for the metabolites have an impact on Cr metabolism [14].

Regulation of L-Arginine Glycine AmidinoTransferase Expression

AGAT reaction is the most likely control step in Cr pathway. Hypothesis supported that Cr exerts a feedback repression of AGAT which most probably serves to conserve the dietary essential amino acids argnine and methionine. Hence, an increase in the serum concentration of Cr due to an endogenous biosyntheses or to dietary Cr supplementation results in decreases of the enzymatic activity of AGAT [15]. The expression of AGAT may be also moderated by dietary and hormonal factors. Hence, Growth Hormone and Cr have an antagonistic action on AGAT expression [16]. AGAT levels are also influenced by dietary intake as dietary deficiency (fasting, protein free diets, vitamine E deficiency) decreases levels of AGAT [17].

Regulation of Cr Transporter Expression

Cr uptake is regulated by a variety of mechanisms. Speer et al. [18] suggested that phosphorylation and glycosylation of the Cr transporter due to changes in the intra or extracellular Cr levels affect regulation of Cr transporter protein. Similarly, Brault et al. [19] demonstrated that variations in Cr uptake rate reflected changes in Cr transporter protein content and activity. Cr transporter protein content is decreased with elevated intra muscular Cr concentrations and would be increased as intra muscular Cr decreases. Cr supplementation decreases Cr transporter expression [20]. Similarly, extracellular Cr down regulates Cr transporter. Na^+ dependent Cr uptake is decreased by extracellular Cr concentration [21]. Insulin increases Na^+K^+ ATP^{ase} activity, which indirectly stimulate Cr transporter activity.

Contribution of Creatine in Physical Exercise

Muscle contraction depends on availability of Adenosine Triphosphate (ATP) and the concomitant discharge of energy [22]. As ATP stored within the cells is very small, cells relay in other regulatory mechanisms to prevent its total degradation. The initial process involves to supply ATP is the high energy PCr, which provides the immediate energy in the beginning of intense or explosive exercise. Hence, it has been reported that PCr is exclusively responsible for ATP regeneration during the initial 10 to 15 seconds of exercise [23]. Also, PCr hydrolysis neither depends on oxygen availability, nor necessitates the achievement of numerous metabolic reactions to buffer energy. In this regard, many activities depend on the phosphagen system such as sprint, team sports, weight lifting and swimming. During short duration and high intensity physical activity, the energy required is almost exclusively provided by ATP and PCr, stored within the muscles [24]. The phosphagen system may continue until depletion of PCr stores [23, 25]. Since, physiologists have addressed Cr supplementation to increase muscles stores of PCr and to exert an ergogenic effect on muscle power output [26] and post exercise PCr resynthesis [27]. Another role attributed to Cr, is in buffering pH to favors ATP resynthesis. Since, Cr helps to prevent acidification and maintain a normal pH of the muscle cells at the onset of maximal exercise [28]. Cr plays an important role in glycolysis regulation. Volek et al. [29,] reported that during intense exercise Cr decreases and phosphofructokinase becomes less inhibited and as a result, the rate of glycolysis increases providing more ATP for muscle.

Effects and Mechanisms of Creatine Supplementation in Sports

Considering the important roles Cr plays, physiologists have worked on Cr supplementation since the early 1900s [2], focusing attention on the benefits for performance. Actually, Cr is probably the most popular ergogenic aid available in the world of athletics. Current research suggests that Cr supplementation increases total Cr content in skeletal muscle, enhances performance during high intensity exercise as well as resistance training [26, 30, 31].

Cr supplementation may improve muscle performance in three different ways; increasing the muscle stores of PCr, for immediate regeneration of ATP in the first few seconds of intense exercise; accelerating PCr resynthesis during recovery periods and depressing the degradation of adenine nucleotides and possibly the accumulation of lactate during exercise [14].

Protocols Applied in Creatine Supplementation

To evaluate the effect of Cr supplementation, two protocols are commonly used. A standard Cr loading protocol (20 g/day for 5 days) is frequently utilized. Generally, this amount of Cr is divided in 5 g/dose, and taken four to five times per day. The loading phase is followed by a maintenance dose of 3-5 g creatine monohydrate/day or 0.03 g creatine monohydrate/kg/day. Another protocol using a dosage of 3 g/day for 28 days was also employed. Most studies have demonstrated an increase in intramuscular Cr levels with both supplement strategies. However, results of these studies were contrivers. This supposes that there are "responders" or "non responders" to Cr supplementation. Recently, Syrotuilik and Bell [32] described individual's biological profile of responders. These authors demonstrated that responders generally have a lower initial quantity of intramuscular Cr and an important proportion of type II fibers, possess a greater fiber cross sectional area and finally have a greater percentage of fat free mass.

Effects of Creatine Supplementation on Physical Performance

It has been reported that Cr supplementation combined with heavy resistance training leads to improve physical performance, fat free mass and muscle morphology [29, 31, 33, 34] (Table 1). It appears that Cr supplementation results in improved in exercise performance by increasing the initial amounts of PCr in the muscle; there is a positive relationship between muscle Cr uptake and exercise performance [35]. Volek et al. [29] reported a significant increase in strength performance in men with resistance training, after 12 weeks of Cr supplementation. The positive effects were attributed to an increased initial amount of PCr in the muscle. Thereby providing a greater initial source of energy and freer Cr aiding the resynthesis rate of PCr during recovery [2].

In elite football players, Cr supplementation (20 g/day for 5 days) combined with resistance training had a positive effect on muscular strength evaluated by a one repetition maximum (1RM) [36].

Table 1. Creatine supplementation and exercise and sport performance summary

Study	dosage	Population	findings
Kreider et al. [79]	20 g of Cr daily for 5 days	25 NCAA American football players	Significant increase in bench press lifting volume and total lifting volume. Improvement of repeated sprint performance Significant increase in body mass (2.4 kg)
Mujika et al [80]	5 g of Cr, four times per day for 6 d	17 highly trained male soccer players	Improvement of repeated sprint performance limited the decay in jumping ability after the intermittent high-intensity exercise
Becque et al [81	Cr Loading: 5g of CrM, four times per day for 5 days. Cr maintenance: 2 g/ day for 5 weeks	Twenty-three male volunteers : (CrM, N = 10; Placebo, N = 13) 6-wk resistance training program	Increases in arm flexor muscular (1RM) Increase fat free mass
Tarnopolsky et al [82]	8 Weeks supplementation with: Group1: 10g Cr + 75g CHO Group 2: 10g PRO + 75g CHO	Young healthy male subjects : (Cr + CHO , N = 11; CHO + PRO, N= 8) 2-month resistance exercise training program	Similar increases in strength greater gains in total mass for the CR-CHO group may have implications for sport-specific performance.
Chrusch et al [83]	Cr Loading: 0.3-g Cr.kg.body weight for 5 days. Maintenance phase: 0.07-g Cr.kg. body weight for 11 weeks	Thirty older men : (Cr, N = 16, or placebo PLA, N = 14), Resistance exercise training program	Improves leg strength, endurance, Improves average power Increases lean tissue mass
Syrotuik et al [84]	Cr loading: 0.3 g/kg/ day for 5 days Maintenance: 0.03 g/kg/day for 5 weeks	Twenty-two rowers: Cr (n = 11) or a placebo (PLA) (n = 12) 6 weeks Resistance exercise training program	No difference between row performance or training volume
Izquierdo et al. [39]	20 g x d(-1) during 5 d	Nineteen trained male handball players	No improvement in upper-body maximal strength and in endurance running performance.
Brose et al [85]	Cr supplementation: group 1: CrM 5 g/d + 2 g of dextrose Group 2 : 7 g of dextrose	Twenty-eight healthy older men and women (group1 : N = 14, group 2, N = 14)	Gains in indices of isometric muscle strength increase in total and fat-free mass,

Table 1. (Continued)

Study	dosage	Population	findings
Volek et al [86]	Cr supplementation: 0.3 g/kg per day CrM for 6 weeks	17 men (CrM: n=9) or placebo (P: n=8) 6 weeks Resistance exercise training program	maintained muscular performance during the initial phase of high-volume resistance training
Ferguson & Syrotuik [87]	10 weeks of CrM supplementation: Cr loading: 0.3 g.kg.body weight for 7 days Cr maintenance: 0.03 g.kg.body weight for 9 weeks	Twenty-six women: Cr (n = 13) or a placebo (PLA) (n = 13) 10 weeks Resistance exercise training program	No difference between groups in strength or lean body mass
Kerksick et al [88]	loading phase: 20 g/day for 5 days maintenance phase: 5 g/day for 23 days	Twenty-four resistance trained males : Group 1 : Cr Group 2: Cr + pinitol (CRP) Group 3: CrM 4 weeks Resistance exercise training program	Improvement in strength and body composition in Cr monohydrate group

NCAA: National Collegiate Athletic Association; Cr: Creatine; PLA: placebo; RM: Repetition Maximum; CHO : carbohydrate; PRO: protein; CrM: Creatine monohydrate; CRP:Creatine pinitol

In accordance, Larson Meuer et al. [37] investigated the acute effects of Cr loading and a periodized resistance program on arm flexor 1RM strength. Fourteen female collegiate soccer players were randomly assigned to either a Cr group (n=7) or a placebo group (n=7). Cr loading consisted in ingesting 7.5 g two times/day for five days, then 5 g/day for 12 weeks.

Following repeated measures and analysis of variance (ANOVA), results indicated that Cr group had significant increases in 1RM strength. Other studies failed to show differences between Cr and placebo group on strength performance. Jakobi et al. [38] reported no effects of Cr supplementation alone upon isometric elbow flexion force, muscle activation and recovery process. These conflicting results can be explained by the possibility that the supplemented group contain a large number of non-responders.

Effects of Creatine Supplementation on Predominantly Anaerobic Exercise

Several studies have examined the effect of Cr loading coupled with resistance training on predominantly anaerobic exercise. These studies have shown a favorable effect of Cr supplementation on muscle performance in short term high intensity exercise. Since, the contribution of PCr hydrolysis to ATP regeneration is relevant in anaerobic exercise [14].

Izquierdo et al. [39] examined the potential ergogenic effects of Cr supplementation on maximal strength and muscle power production during repetitive high-power-output exercise bouts (MRPB) as well as repeated running sprints in handball players. Nineteen trained male handball players were randomly assigned in a double-blind fashion to either Cr (n = 9) or placebo (n = 10) group. This study showed that short term Cr administration had a significant ergogenic effect upon maximal repetitive upper and lower body high power exercise bouts, and total repetition performed to fatigue, as well as repeated sprint.

One of the potentially most beneficial effects of Cr supplementation in power athletes is an increase of the amount of work performed during a series of maximal effort muscle contraction. Volek et al. [40] reported that Cr supplementation (25 g/day for 7 days) resulted in significant improvement in exercise performance during five sets of bench press and jump squats in comparison to a placebo group. Cr supplementation resulted in a significant increase in repetitions performed during bench press for set 2, while peak power output significantly increased in the jump squat during set 5. The authors concluded that increase in exercise performance and body mass (1.3 kg) associated with one week of Cr supplementation are not due to any measurable alteration in circulating concentrations of steroid hormones, as pre- and post-exercise testosterone and cortisol values did not differ significantly between groups. Additionally, when Cr supplementation was extended for a further 11 days [41], significant increases were recorded in all 5 sets of bench press repetitions (26.6%) and jump squat peak power output (4.7%). Improvements are most likely related to an increase in energy substrate availability and resynthesis. Moreover, Vandenberghe et al. [42] indicate that Cr supplementation (20 g/day for 4 days) increased muscle PCr concentration by 6%. This increase was maintained thereafter, during 10 weeks of training associated with low-dose Cr intake (5 g/day). Compared with placebo, maximal strength of the muscle groups trained, maximal intermittent exercise capacity of the arm flexors, and fat-free mass were increased 20-25%, 10-25%, and 60% more, respectively, during Cr supplementation.

Effects of Creatine Supplementation on Predominantly Aerobic Exercise

In terms of sport performance, Cr supplementation has been shown to be more effective on predominantly anaerobic exercise with short duration (less than 3 minutes). However, there is some evidence of the potential positive ergogenic effects on endurance activities. Chwalbińska-Moneta et al. [43] investigated the effect of oral Cr supplementation on aerobic performance in 16 elite male rowers after consuming 20 g Cr/day for five days; a significant decrease resulted in blood lactate accumulation during endurance training. It appears that Cr supplementation improves endurance expressed by increase in lactate threshold. Rico and Marco [44] examined the effect of Cr supplementation on performance in cyclists during alternating intensity exercise as 30% and 90% of maximal power output. Fourteen male subjects were randomly assigned to either Cr group (n=7; 20 g/day for 5 days) placebo group (n=7). Time of exhausting only was improved in Cr group (29.9±3.8 vs 36.5±5.7). Authors reported an increase in oxygen consumption following Cr ingestion when cyclists cycled at 90% of maximal power output, but lower submaximal oxygen consumption when cyclists cycled at 60% of maximal power output. It appeared that Cr supplementation improves performance by enhancing oxidative phosphorylation at alternating intensity exercise. In another study involving a short loading Cr protocol (20 g/day for 6 days), Balsom et al. [45] reported no enhancement in performance time of a 6 km cross country run. The lack of effect of Cr supplementation on performance in the present study may be due to the increase of body mass associated with Cr ingestion. Manjarrez-Montes de Oca et al. [46] stated that Cr supplementation may increase fat mass and serum triglycerides concentration in young male TKD practitioners without improvement in anaerobic power. Cr supplementation appears to be safe, but athletes should be careful when they seek fat loss.

High-Intensity Performance and Cr Supplementation

Using a double-blind, placebo-control design involving 32 elite male and female swimmers from the Australian National Team, Burke et al. [47] reported that Cr supplementation (20 g/day for 5 days) did not enhance performance in maximal single effort swim sprints of 25 m, 50 m, and 100 m, each interspersed with 10 min recovery period. Given the length of the recovery period, resynthesis of ATP would be complete without Cr supplementation; therefore, an increase in performance would not be expected.

In a similar study, Mujika et al. [48] assigned 20 male and female swimmers in a randomized, double-blind manner to either Cr supplementation (20 g/day for 5 days) or placebo groups in order to investigate the effect on 25 m, 50 m, and 100 m swim sprint performance. They reported no performance differences between the groups, however, a significant increase in body weight was found in the Cr supplemented group. The authors suggested that the increase in body weight experienced by subjects following Cr supplementation could result in a concomitant increase in drag force and altered stroke mechanics. Such a mechanism is a likely reason why no ergogenic effect was present. Performing 30 s maximal cycling (Wingate) task after Cr supplementation (20 g/day for 3 days), Odland et al. [49] found that Cr supplementation neither increased resting muscle PCr, nor affected the single short term maximal cycling performance. The most likely explanation for this is that the increase in muscle PCr content after Cr supplementation was insufficient to induce an enhanced sprint performance and to allow an improved rate of PCr resynthesis after exercise. Alternatively, it is also possible that Cr supplementation does not enhance sprint performance during brief maximal exercise. Following this line of investigation, Snow et al. [50] utilized a double-blind crossover design on untrained men performing 120 s maximal sprint on an cycle ergometer after Cr supplementation (30 g/day for 5 days). The data demonstrated that Cr supplementation increased muscle PCr content, but the increase did not induce an improved sprint exercise performance or alterations in anaerobic muscle metabolism. In conclusion, the authors reported that a small, yet significant increase in muscle PCr content occurred but this increase, however, did not result in an improved sprint-exercise performance or any alterations in markers of muscle anaerobic energy metabolism during, and in recovery from sprint exercise.

Effects of Creatine Supplementation on Glycogen Stores

Cr supplementation may also enhance glycogen accumulation and GLUT4 expression when it is combined with a glycogen depletion exercise. Hickner el al. [51] found that 28 days of Cr supplementation at 3 g/day enhanced initial maintained a higher level of muscle glycogen during 2 hours of cycling. Op't Eijnde et al. [52] confirmed that Cr supplementation combined with a high carbohydrate regime have a potential and benefit of heightened muscle glycogen stores.

Effects of Creatine Supplementation on Body Mass

Several studies reported changes in body mass following Cr supplementation. The average increase reported in the literature amounts to 1 to 2 kg of total body mass. The increase has been attributed to fat free mass (Table 2).

Table 2. Body mass changes induced by Creatine supplementation

Study	Dosage	Population	Effect on body mass
Aguiar et al [89]	Cr: 5g/day for 12 weeks	Eighteen Older women (Cr, N = 9, or placebo PLA, N =9) 12 weeks Resistance exercise training program	Significant Increase in body mass: + 3.2 FFM + 2.8 MM
Sculthorpe et al [90]	Cr loading: 25g/day for 5 days Maintenance: 5g/day for 3 days	Forty Active men (Cr, N=20, or placebo, N = 20)	Significant Increase in body masse +1.6 kg
Rawson et al [59]	C r: 0.03g/kg/day for 6 weeks	Twenty healthy men and women (Cr, N= 10, 4women or placebo, N= 10, 4 women) 6weeks resistance exercise training program	No differences between groups
Cancela et al [91]	Cr loading: 15g/day for 7 days Maintenance: 3g/day for 49 days	Forteen football players (Cr, N = 7, or placebo PLA, N =7)	Increase on total body mass
Chilibeck et al [60]	Cr: 0.1g/kg/day for 8 weeks	Eighteen rugby football players 8weeks of a season of a rugby union football	No differences between groups
Branch et al [61]	20g/day for 28 days	Competitive cyclists and triathletes	No differences between groups
Gotshalk et al [92]	C r: 0.3g/kg/day for 7 days	Thirty old women	Increase in body mass
Wright et al [93]	Cr:20g/day for 6 days	Ten active men	Increase in body mass
Glaister et al [94]	Cr:20g/day for 6 days	42 physical active men	Increase in body mass Decrease in fat mass

Cr: creatine; PLA: placebo: FFM:fat free mass, MM:muscle mass.

Changes in body weight observed following Cr ingestion was explained by an increase in body water, especially in the intracellular compartments [10].

Hultman et al. [53] observed a 0.6 liter decline in urinary volume after acute Cr loading. They suggested that increase in body mass can probably be attributed to body water retention. Furthermore, Zeigenfuss et al [54] reported an increase in thigh skeletal muscle volume, total body mass and intra cellular water volume in aerobic and cross trained men following acute Cr consumption. Francaux and Poortmans [55] reported also an increase in body mass in volunteers consuming Cr supplements over 9 weeks. Authors suggested that the increase in body mass (measured by bio impedance) was attributed to an increase in body water content and more specifically to an increase in the volume of the intracellular compartment. Bessman and Savabi [56] reported that Cr could induce muscle hypertrophy in adult. Since, they suggest that there is an increased uptake of amino acids following Cr supplementation. In addition, an enhancement of biosynthesis of myofrillar proteins has occurred after Cr supplementation. In accordance, Cribb et al. [57] showed an improvement on lean body mass, fiber cross sectional area and contractile protein in trained young males following a multi nutrient supplement (0.1 g/kg/day of creatine, 1.5 g/kg/day of protein and carbohydrate) combined with resistance training. It appeared that Cr supplementation had a positive effect on skeletal muscle hypertrophy. Since, Deldicque et al [58] demonstrated an increase in collagen mRNA, glucose transporter 4 (GLUT 4) and myosine heavy chain IIA after acute Cr loading (21 g/d for 5 days). These authors suggested that short term Cr supplementation combined with resistance training induced changes in gene expression promoting an anabolic environment.

It should not be ignored that conflicting reports suggested no changes in body mass after Cr supplementation [59-61]. Several explanations can be put forward for the discrepant findings, which, however, will need further investigation such as, too short or too long a period of Cr supplementation, insufficient daily Cr dosage, or a high proportion of non responders among the subjects analyzed.

Potential Adverse Effects of Creatine Supplementation

Cr supplementation is generally considered safe and well-tolerated by humans and animals [62]. For instance, no significant adverse effects have been reported in studies of patients with inborn error of Cr synthesis given up to 0.8 g/kg daily for two years [63]. Likewise, no major health issues were reported in a study of 9 healthy athletes administered 1–20 g/day from 1 to 4 times/day for up to five years [64] Nevertheless, empirical and anecdotal accounts do exist that describe mild to moderate side effects of daily Cr supplementation.

The majority reported side effects of weight gain, gastrointestinal distress, altered insulin production, inhibition of endogenous Cr synthesis, renal dysfunction, or dehydration. Fewer have reported disturbances in mood and anxiety. Experts generally agree that there is sufficient evidence to confirm that 5 g/day of Cr is generally harmless to healthy adults, although there is not enough evidence to make an informed recommendation in favor or against doses higher than 5 g/day [65].

Gastrointestinal Distress

Sufficient evidence has been documented to support gastrointestinal distress as a consequence of daily Cr supplementation. Mild to moderate gastrointestinal issues have been reported in a number of experimental and clinical trials, including diarrhea, nausea, and/or vomiting [66, 67]. In addition, abdominal discomfort, increased or decreased appetite, weight gain, stomach distress and loose stool have been reported [68, 69].

Renal Dysfunction

While some case studies have suggested that Cr causes renal dysfunction, most empirical studies in humans and animals indicate that it is more probable that Cr worsens pre-existing renal disease [70, 71]. In other words, it is unlikely that healthy adults without a history of renal disease will develop kidney problems as a consequence of Cr supplementation unless there are other exacerbating factors involved, such as use of illicit anabolic–androgenic steroids, non-steroidal anti-inflammatory agents, nephrotoxic or renal-cleared medications, or diuretics.

It has also been argued that Cr supplementation can confound renal analyses because serum creatinine is the most widely used marker of renal function [72]. Cr supplementation increases levels of creatinine, which can be falsely interpreted as an indication of renal dysfunction because most laboratories factor in serum creatinine levels when estimating glomerular filtration rate. While research is ongoing, the majority of studies conclude that Cr supplementation is generally not harmful to the kidneys when used as directed [73].

Body Weight and Water Retention

Increased body weight has been reported in humans following Cr supplementation. This is most probably due to the fact that Cr is an osmotically active substance that increases cell water retention inside cells [10]. The water retained is lost following discontinuation of Cr supplementation.

Dehydration

Few empirical studies using control groups and blinding have documented dehydration as a side effect of Cr supplementation [74]. To be on the safe side, physicians typically advise drinking extra water and avoiding caffeine when taking Cr supplements [75].

Mood and Anxiety

Cr supplementation can increase the risk of mania or depression in susceptible individuals. It is also possible that long-term high dosing of Cr alters Cr transporter function or Cr kinase activity in a manner that adversely affects emotional regulation [76]. Further research is required before drawing definitive conclusions, but caution is warranted for individuals at risk. Despite the current popularity of Cr supplementation, it must be kept in mind that it is virtually impossible to ingest 20–30 g Cr/day. Based on the lack of proper investigations on the potential side effects of this compound and on its mechanism of action, discussions should continue on whether Cr supplementation is a legal ergogenic aid or whether it should be regarded as a doping strategy [77, 78].

Conclusion

Cr supplementation has positive effects on boosting the effects of resistance training for improving strength and performance. It seems also to produce positive effects on enhancement of body composition by increasing fat free mass. Regarding anaerobic performance, Cr supplementation seems to be useful to attenuate fatigue symptoms over multiple bouts of high intensity short duration exercises. For aerobic endurance performance, Cr supplementation seems to increase plasma volume, glycogen stores and ventilator threshold and to decrease oxygen consumption in sub maximal exercise. In this regard, Cr may be an effective ergogenic aid for athletes.

Despite the increase of oral Cr consumption among professional and amateur athletes, few studies emphasis potential adverse effects of Cr supplementation. The available evidence indicates that oral Cr consumption at the correct dose appears to be safe for athletes engaged in intense training and competition. Yet, it must be used with precaution in individuals at risk. It is advisable to monitor liver and kidney function in individuals consuming Cr supplements; regular checkups are necessary to detect any potential dysfunction. In addition, long term and epidemiological data should be elaborated to determine the safety of Cr in healthy individuals. More rigorous clinical research need to be carried out, and the pharmacokinetics and dose–response effects in athletes need to be more thoroughly investigated.

References

[1]   Needham DM. Machina carnis: the biochemistry of muscular contraction in its historical development. Cambridge: Cambridge University Press, 1971.

[2]   Demant TW, Rhodes EC. Effects of creatine supplementation on exercise performance. *Sports Med* 1999;28:49-60.

[3]   Terjung RL, Clarkson P, Eichner ER, Greenhaff PL, Hespel PJ, Israel RG, Kraemer WJ, Meyer RA, Spriet LL, Tarnopolsky MA, Wagenmakers AJ, Williams MH. American College of Sports Medicine roundtable. The physiological and health effects of oral creatine supplementation. *Med. Sci. Sports Exerc* 2000;32:706-17.

[4]   Pastoris O, Boschi F, Verri M, Baiardi P, Felzani G, Vecchiet J, Dossena M, Catapano M. The effects of aging on enzyme activities and metabolite concentrations in skeletal muscle from sedentary male and female subjects. *Exp. Gerontol* 2000;35:95-104.

[5]   Rawson ES, Gunn B, Clarkson PM.The effects of creatine supplementation on exercise-induced muscle damage. *J. Strength Cond Res*. 2001;15:178-84.

[6]   Silber ML. Scientific facts behind creatine monohydrate as sport nutrition supplement. *J. Sports Med Phys Fitness* 1999;39:179-88.

[7]   Devlin TM. *Textbook of biochemistry: with clinical correlations*. New York: Wiley-Liss, 1992.

[8]   Harris RC, Soderlund K, Hultman E. Elevation of creatine in resting and exercised muscle of normal subjects by creatine supplementation. *Clin. Sci.* 1992; 83: 367-74.

[9]   Schoch RD, Willoughby D, Greenwood M. The regulation and expression of the creatine transporter: a brief review of creatine supplementation in humans and animals. *J. Int Soc Sports Nutr.* 2006;3:60-6.

[10]   Bemben M, Lamont H: Creatine supplementation and exercise performance: recent findings. *Sports Med* 2005, 35:107–25.

[11]   Brosnan JT, da Silva RP, Brosnan ME: The metabolic burden of creatine synthesis. *Amino Acids* 2011, 40:1325–31.

[12]   Söderlund K, Hultman E. ATP and phosphocreatine changes in single human muscle fibers after intense electrical stimulation. *Am. J. Physiol.* 1991; 261:E737-41.

[13]   Greenhaff PL, Nevill ME, Soderlund K, Bodin K, Boobis LH, Williams C, Hultman E. The metabolic responses of human type I and II muscle fibres during maximal treadmill sprinting. *J. Physiol.* 1994; 478:149-55.

[14]   Wyss M, Kaddurah-Daouk R. Creatine and creatinine metabolism. *Physiol. Rev.* 2000;80:1107-213.

[15]   Dinning JS, Day PL. Creatinuria during recovery from aminopterin-induced folic acid deficiency in the monkey. *J. Biol. Chem* 1949;181:897-903.

[16]   Guthmiller P, Van Pilsum JF, Boen JR, McGuire DM. Cloning and sequencing of rat kidney L-arginine:glycine amidinotransferase. Studies on the mechanism of regulation by growth hormone and creatine. *J. Biol. Chem.* 1994;269:17556-60.

[17]   Van Pilsum JF, McGuire DM, Towle H. On the mechanism of the alterations of rat kidney transamidinase activities by diet and hormones. *Adv. Exp. Med. Biol.* 1982;153:291-7.

[18]   Speer O, Neukomm LJ, Murphy RM, Zanolla E, Schlattner U, Henry H, Snow RJ, Wallimann T. Creatine transporters: a reappraisal. *Mol. Cell Biochem.* 2004;256-257:407-24.

[19]   Brault JJ, Terjung RL. Creatine uptake and creatine transporter expression among rat skeletal muscle fiber types. *Am. J. Physiol Cell Physiol.* 2003;284:1481-9.

[20]   Guerrero-Ontiveros ML, Wallimann T. Creatine supplementation in health and disease. Effects of chronic creatine ingestion in vivo: down-regulation of the expression of creatine transporter isoforms in skeletal muscle. *Mol. Cell Biochem.* 1998;184:427-37.

[21] Loike JD, Zalutsky DL, Kaback E, Miranda AF, Silverstein SC. Extracellular creatine regulates creatine transport in rat and human muscle cells. *Proc. Natl. Acad. Sci. USA* 1988;85:807-11.

[22] Glaister M. Multiple sprint work : physiological responses, mechanisms of fatigue and the influence of aerobic fitness. *Sports Med* 2005;35:757-77.

[23] Bogdanis GC, Nevill ME, Boobis LH, Lakomy HK. Contribution of phosphocreatine and aerobic metabolism to energy supply during repeated sprint exercise. *J. Appl. Physiol* 1996;80:876-84.

[24] McArdle WD, Katch FI, Katch VL. *Exercise physiology: energy, nutrition and human performance.* Philadelphia: Lea &Febiger, 1991

[25] Withers RT, Sherman WM, Clark DG, Esselbach PC, Nolan SR, Mackay MH, Brinkman M. Muscle metabolism during 30, 60 and 90 s of maximal cycling on an air-braked ergometer. *Eur. J. Appl. Physiol Occup. Physiol.* 1991;63:354-62.

[26] Casey A, Constantin-Teodosiu D, Howell S, et al. Creatine ingestion favorably affects performance and muscle metabolism during maximal exercise in humans. *Am. J. Physiol* 1996;271:E31-7.

[27] Greenhaff PL, Bodin K, Soderlund K, Hultman E. Effect of oral creatine supplementation on skeletal muscle phosphocreatine resynthesis. *Am. J. Physiol.* 1994;266:E725-30.

[28] Forsberg AM, Nilsson E, Werneman J, Bergström J, Hultman E.Muscle composition in relation to age and sex. *Clin. Sci.* [Lond] 1991;81: 249-56.

[29] Volek JS, Duncan ND, Mazzetti SA, Staron RS, Putukian M, Gómez AL, Pearson DR, Fink WJ, Kraemer WJ.Performance and muscle fiber adaptations to creatine supplementation and heavy resistance training. *Med. Sci. Sports Exerc* 1999;31:1147-56.

[30] Grindstaff PD, Kreider R, Bishop R, Wilson M, Wood L, Alexander C, Almada A. Effects of creatine supplementation on repetitive sprint performance and body composition in competitive swimmers. *Int. J. Sport Nutr.* 1997;7:330-46.

[31] Kreider RB. Effects of creatine supplementation on performance and training adaptations. *Mol. Cell Biochem.* 2003;244:89-94.

[32] Syrotuik DG, Bell GJ. Acute creatine monohydrate supplementation: Adescriptive physiological profile of responders vs. nonresponders. *J. Strength Cond Res* 2004;18:610-7.

[33] Dempsey R, Mazzone M, Meurer L: Does oral creatine supplementation improve strength? A meta-analysis. *J. Fam Pract* 2002;51:945–51.

[34] Volek J, Rawson E: Scientific basis and practical aspects of creatine supplementation for athletes. *Nutrition* 2004;20:609–14.

[35] Casey A, Greenhaff P: Does dietary creatine supplementation play a role in skeletal muscle metabolism and performance? *Am. J. Clin. Nutr* 2000;72:607S–17S.

[36] Bemben MG, Bemben DA, Loftiss DD, Knehans AW. Creatine supplementation during resistance training in college football athletes. *Med. Sci Sports Exerc.* 2001 ;33:1667-73.

[37] Larson-Meyer DE, Hunter GR, Trowbridge CAThe effect of creatine supplementation on muscle strength and body composition during off-season training in female soccer player. *J. Strength Cond Res* 2000;14: 434-42.

[38] Jakobi J, Rice C, Curtin S, Marsh G: Contractile properties, fatigue and recovery are not influenced by short-term creatine supplementation in human muscle. *Exp. Physiol.* 2000;85:451–60.

[39] Izquierdo M, Ibañez J, González-Badillo JJ, Gorostiaga EM. Effects of creatine supplementation on muscle power, endurance, and sprint performance. *Med. Sci Sports Exerc* 2002;34:332-43.

[40] Volek, J.S., Boetes, M., Bush, J.A., Putukian, M., Sebastianelli, W.J. and Kraemer, W.J. Response of testosterone and cortisol concentrations to high-intensity resistance exercise following creatine supplementation. *Journal of Strength and Conditioning Research* 1997a;11:182-7.

[41] Volek, J.S., Kraemer, W.J., Bush, J.A., Boetes, M., Incledon, T., Clark, K.L. and Lynch, J.M. Creatine supplementation enhances muscular performance during high-intensity resistance exercise. *Journal of the American Dietetic Association* 1997b;97: 765-70.

[42] Vandenberghe, K., Goris, M., Van Hecke, P.M., Leemputte, V., Vangerven, L. and Hespel, P. Long-term creatine intake is beneficial to muscle performance during resistance training. *Journal of Applied Physiology* 1997;83:2055-63.

[43] Chwalbiñska-Moneta J.Effect of creatine supplementation on aerobic performance and anaerobic capacity in elite rowers in the course of endurance training. *Int. J. Sport Nutr Exerc Metab.* 2003;13:173-83.

[44] Rico-Sanz J, Marco MTM. Creatine enhances oxygen uptake and performance during alternating intensity exercise. *Med. Sci Sports Exerc* 2000;32:379-85.

[45] Balsom PD, Harridge SDR, Soderlund K. Creatine supplementation per se does not enhance endurance exercise performance. *Acta Physiol Scand.* 1993;149:521-3.

148   Ayed Ikram Bezrati-Ben, Fehmi Nasrallah, Raouf Hammami et al.

[46] Manjarrez-Montes de Oca R, Farfán-González F, Camarillo-Romero S, Tlatempa-Sotelo P, Francisco-Argüelles C, Kormanowski A, González-Gallego J, Alvear-Ordenes I. Effects of creatine supplementation in taekwondo practitioners. *Nutr. Hosp* 2013;28:391-9.

[47] Burke LM, Pyne DB, Telford RD. Effect of oral creatine supplementation on single-effort sprint performance in elite swimmers. *Int. J. Sport Nutr.* 1996;6:222-33

[48] Mujika I, Chatard JC, Lacoste L, Barale F, Geyssant A.Creatine supplementation does not improve sprint performance in competitive swimmers. *Med. Sci. Sports Exerc*. 1996;28:1435-41.

[49] Odland LM, MacDougall JD, Tarnopolsky MA, Elorriaga A, Borgmann A Effect of oral creatine supplementation on muscle [PCr] and short-term maximum power output. *Med. Sci Sports Exerc* 1997;29:216-9.

[50] Snow RJ, McKenna MJ, Selig SE, Kemp J, Stathis CG, Zhao S. Effect of creatine supplementation on sprint exercise performance and muscle metabolism. *J. Appl. Physiol* 1998;84:1667-73.

[51] Hickner R, Dyck D, Sklar J, Hatley H, Byrd P: Effect of 28 days of creatine ingestion on muscle metabolism and performance of a simulated cycling road race. *J. Int Soc Sports Nutr.* 2010;7:26.

[52] Op 't Eijnde B, Urso B, Richter EA, Greenhaff PL, Hespel P: Effect of oral creatine supplementation on human muscle GLUT4 protein content after immobilization. *Diabetes* 2001;50:18–23.

[53] Hultman E, Söderlund K, Timmons JA, Cederblad G, Greenhaff PL. Muscle creatine loading in men. *J. Appl. Physiol.* 1996;81:232-7.

[54] Ziegenfuss T, Lemon P, Rogers M, et al. Acute creatine ingestion: effects on muscle volume, anaerobic power, fluid volumes and protein turnover [abstract]. *Med. Sci. Sports Exerc* 1997;29:S127

[55] Francaux M, Poortmans JR. Effects of training and creatine supplement on muscle strength and body mass. *Eur. J. Appl Physiol Occup Physiol* 1999;80:165-8.

[56] Bessman S, Savabi F. The role of the phosphocreatine energy shuttle in exercise and muscle hypertrophy. In: Taylor A, Gollnick P, Green H, et al., editors. *Biochemistry of exercise* VII. Champaign [IL]: Human Kinetics, 1990:167-78.

[57] Cribb PJ, Williams AD, Hayes A. A creatine-protein-carbohydrate supplement enhances responses to resistance training. *Med. Sci. Sports Exerc.* 2007;39:1960-8.

[58] Deldicque L, Atherton P, Patel R, Theisen D, Nielens H, Rennie MJ, Francaux M. Effects of resistance exercise with and without creatine

supplementation on gene expression and cell signaling in human skeletal muscle. *J. Appl. Physiol* 2008;104:371-8.

[59] Rawson ES, Stec MJ, Frederickson SJ, Miles MP. Low-dose creatine supplementation enhances fatigue resistance in the absence of weight gain. *Nutrition* 2011;27:451-5.

[60] Chilibeck PD, Magnus C, Anderson M. Effect of in-season creatine supplementation on body composition and performance in rugby union football players. *Appl. Physiol Nutr Metab* 2007;32:1052-7.

[61] Branch JD, Schwarz WD, Van Lunen B. Effect of creatine supplementation on cycle ergometer exercise in a hyperthermic environment. *J.Strength Cond Res* 2007;21:57-61.

[62] Rodriguez NR, Di Marco NM, Langley S. American college of sports medicine position stand. Nutrition and athletic performance. *Med. Sci. Sports Exerc.* 2009;41: 709–31.

[63] Braissant O, Henry H, Villard AM, Speer O, Wallimann T, Bachmann C. Creatine synthesis and transport during rat embryogenesis: spatiotemporal expression of AGAT, GAMT and CT1. BMC Dev Biol 2005;5:9.

[64] Poortmans JR, Francaux M. Long-term oral creatine supplementation does not impair renal function in healthy athletes. *Med. Sci. Sports Exerc.* 1999;31:1108–10.

[65] Shao A, Hathcock JN. Risk assessment for creatine monohydrate. *Regul. Toxicol. Pharmacol.* 2006;45:242–51.

[66] Groeneveld GJ, Veldink JH, van der Tweel I, Kalmijn S, Beijer C, de Visser M, Wokke JH, Franssen H, van den Berg LH. A randomized sequential trial of creatine in amyotrophic lateral sclerosis. *Ann. Neurol* 2003;53:437-45.

[67] Astorino TA, Marrocco AC, Gross SM, Johnson DL, Brazil CM, Icenhower ME, Kneessi RJ Is running performance enhanced with creatine serum ingestion? *J. Strength Cond Res* 2005;19:730-4.

[68] Kendall RW, Jacquemin G, Frost R, Burns SP. Creatine supplementation for weak muscles in persons with chronic tetraplegia: a randomized double-blind placebo-controlled crossover trial. *J. Spinal Cord. Med.* 2005;28:208-13.

[69] Tarnopolsky MA.Potential benefits of creatine monohydrate supplementation in the elderly. *Curr. Opin. Clin. Nutr. Metab Care* 2000;3:497-502.

[70] Sheth NP, Sennett B, Berns JS. Rhabdomyolysis and acute renal failure following arthroscopic knee surgery in a college football player taking creatine supplements. *Clin. Nephrol* 2006;65:134-7.

[71] Thorsteinsdottir B, Grande JP, Garovic VD. Acute renal failure in a young weight lifter taking multiple food supplements, including creatine monohydrate. *J. Ren. Nutr* 2006;16:341-5.

[72] Gualano B, Ugrinowitsch C, Novaes RB, Artioli GG, Shimizu MH, Seguro AC, Harris RC, Lancha AH Jr Effects of creatine supplementation on renal function: a randomized, double-blind, placebo-controlled clinical trial. *Eur. J. Appl. Physiol* 2008;103:33-40.

[73] Dalbo VJ, Roberts MD, Stout JR, Kerksick CM. Putting to rest the myth of creatine supplementation leading to muscle cramps and dehydration. *Br. J. Sports Med.* 2008;42:567-73.

[74] Juhn MS, O'Kane JW, Vinci DM. Oral creatine supplementation in male collegiate athletes: a survey of dosing habits and side effects. *J. Am. Diet Assoc* 1999;99:593-5.

[75] Vahedi K, Domigo V, Amarenco P, Bousser MG. Ischaemic stroke in a sportsman who consumed MaHuang extract and creatine monohydrate for body building. *J. Neurol Neurosurg Psychiatry* 2000;68:112-3.

[76] Allen PJ. Creatine metabolism and psychiatric disorders: Does creatine supplementation have therapeutic value? *Neurosci. Biobehav. Rev.*2012;36:1442-62.

[77] Masood E. Performance-enhancers pose dilemma for rule-makers. Nature 1996;382:16.

[78] Williams MH, Branch JD. Creatine supplementation and exercise performance: an update. *J. Am. Coll Nutr.* 1998;17:216-34.

[79] Kreider RB, Ferreira M, Wilson M, Grindstaff P, Plisk S, Reinardy J, Cantler E, Almada AL. Effects of creatine supplementation on body composition, strength, and sprint performance. *Med. Sci. Sports Exerc* 1998;30:73-82.

[80] Mujika I, Padilla S, Ibañez J, Izquierdo M, Gorostiaga E. Creatine supplementation and sprint performance in soccer players. *Med. Sci. Sports Exerc.* 2000;32:518-25.

[81] Becque MD, Lochmann JD, Melrose DR.Effects of oral creatine supplementation on muscular strength and body composition. *Med. Sci. Sports Exerc* 2000;32:654-8.

[82] Tarnopolsky MA, Parise G, Yardley NJ, Ballantyne CS, Olatinji S, Phillips SM.Creatine-dextrose and protein-dextrose induce similar strength gains during training. *Med. Sci. Sports Exerc* 2001;33:2044-52.

[83] Chrusch MJ, Chilibeck PD, Chad KE, Davison KS, Burke DG. Creatine supplementation combined with resistance training in older men. *Med. Sci. Sports Exerc.* 2001;33:2111-7.

[84] Syrotuik DG, Game AB, Gillies EM, Bell GJ. Effects of creatine monohydrate supplementation during combined strength and high intensity rowing training on performance. *Can. J. Appl. Physiol* 2001;26:527-42.

[85] Brose A, Parise G, Tarnopolsky MA.Creatine supplementation enhances isometric strength and body composition improvements following strength exercise training in older adults. *J. Gerontol. A Biol. Sci. Med. Sci.* 2003;58:11-9.

[86] Volek JS, Ratamess NA, Rubin MR, Gómez AL, French DN, McGuigan MM, Scheett TP, Sharman MJ, Häkkinen K, Kraemer WJ.The effects of creatine supplementation on muscular performance and body composition responses to short-term resistance training overreaching. *Eur. J. Appl. Physiol.* 2004;91:628-37.

[87] Ferguson TB, Syrotuik DG. Effects of creatine monohydrate supplementation on body composition and strength indices in experienced resistance trained women. *J. Strength Cond Res.* 2006;20:939-46.

[88] Kerksick CM, Wilborn CD, Campbell WI, Harvey TM, Marcello BM, Roberts MD, Parker AG, Byars AG, Greenwood LD, Almada AL, Kreider RB, Greenwood M. The effects of creatine monohydrate supplementation with and without D-pinitol on resistance training adaptations. *J. Strength Cond Res.* 2009;23:2673-82.

[89] Aguiar AF, Januário RS, Junior RP, Gerage AM, Pina FL, do Nascimento MA, Padovani CR, Cyrino ES. Long-term creatine supplementation improves muscular performance during resistance training in older women. *Appl. Physiol Nutr. Metab.* 2010;35:507-11.

[90] Sculthorpe N, Grace F, Jones P, Fletcher I. The effect of short-term creatine loading on active range of movement. *Appl. Physiol. Nutr. Metab* 2010;35:507-11.

[91] Cancela P, Ohanian C, Cuitiño E, Hackney AC. Creatine supplementation does not affect clinical health markers in football players. *Br. J. Sports Med.* 2008;42:731-5.

[92] Gotshalk LA, Kraemer WJ, Mendonca MA, Vingren JL, Kenny AM, Spiering BA, Hatfield DL, Fragala MS, Volek JS. Creatine supplementation improves muscular performance in older women. *Eur. J. Appl. Physiol.* 2008;102:223-31.

[93] Wright GA, Grandjean PW, Pascoe DD. The effects of creatine loading on thermoregulation and intermittent sprint exercise performance in a hot humid environment. *J. Strength Cond Res* 2007;21:655-60.

[94] Glaister M, Lockey RA, Abraham CS, Staerck A, Goodwin JE, McInnes G. Creatine supplementation and multiple sprint running performance. *J. Strength Cond Res*. 2006;20:273-7.

Index

B

C

F

G

H

I

N

O